INTRODUCTION TO

Abstract Algebra

BY ROY DUBISCH

The Teaching of Mathematics

Intermediate Algebra
with Vernon E. Howes and Steven J. Bryant

Self-Teaching Intermediate Algebra
with Vernon E. Howes

Higher Algebra for the Undergraduate
with Marie J. Weiss

INTRODUCTION TO
Abstract Algebra

ROY DUBISCH

PROFESSOR OF MATHEMATICS
UNIVERSITY OF WASHINGTON

JOHN WILEY & SONS, INC.
New York London Sydney

Library of Congress Catalog Card Number: 65-16405
Printed in the United States of America

Preface

The aim of this text is to provide a gradual introduction to the basic concepts of abstract algebra. It begins very slowly with the development of a set of postulates for the natural numbers and, throughout, alternates between discussions of familiar systems and their more abstract generalizations. Thus the discussion of natural numbers and the integers is followed by a discussion of integral domains; the discussion of rational numbers is followed by a discussion of groups and fields, etc.; and in the last two chapters the concept of a vector is applied to polynomials, and standard results on the theory of equations are presented. Following the same principle of gradual development, the text presents induction in Chapter 2 in its simplest form and then returns to a more sophisticated treatment in Chapter 5; the simple concept of a group is presented in Chapter 7 with further work on this topic given in Chapter 9.

In addition to a large number of problems of varying degrees of difficulty, the text includes, at the end of each chapter, a list of problems for further investigation. These are, in general, open-ended questions designed to stimulate active investigation by the student and to lead him to further reading not only in the standard texts on the subject but also in such journals as the *American Mathematical Monthly*.

For a student with a background of four years of conventional secondary school mathematics, the material in this text should serve for approximately two-thirds of a year's course in abstract algebra to be followed by material in linear algebra. Students with a background of more modern secondary school mathematics should be able to cover the material in less time.

Within any chapter, theorems and definitions are numbered sequentially. A reference to a theorem or definition in another chapter will include the chapter number and the number of the theorem or definition. Thus, for example, Theorem 7.2 refers to Theorem 2 of Chapter 7. An asterisk preceding a problem indicates a more difficult problem; a notation such as [B1] refers to the books listed in the bibliography; a list of symbols used appears just before the text.

ROY DUBISCH

Seattle, Washington
February, 1965

Contents

Chapter 6 (*continued*)

Symbols used and pages where first introduced

Symbol	Meaning	Page
$\in$	set membership	2
$\subset, \subseteq$	set inclusion	2
$\emptyset$	null set	1
$\cup$	Union for sets	3
$\cap$	intersection for sets	3
$\bar{A}$	complement of a set	4
A/B	for sets	4
$A \times B$	Cartesian product of sets	5
N	natural numbers	17
$\sim$	equivalence	18
$I\langle a, b\rangle$	integer	31
I	integers	38
R	rational numbers	38
R^*	real numbers	38
C	complex numbers	38
$x \mid y$	x divides y	38
C_i	residue class	39
J_m	set of residue classes	39
$>$	greater than	47
$<$	less than	47
$\|a\|$	absolute value	48
$R\langle a, b\rangle$	rational number	60
S_n	symmetric group	78
A_n	alternating group	82
$\cong$	isomorphic	85
$[a]$	greatest integer	91
$M(S)$	set of 1–1 mappings	92
R^+	positive rational numbers	105
$\bar{s}$	residue class	129
$S–R$, S/R	residue class ring	130
K_σ	kernel of homomorphism	133
$H \Delta G$	normal subgroup	138
cis θ	complex number	148
$V_n(F)$	space of n-tuples	155

1

Sets

1. Definition

Modern algebra, as we shall see, does not deal only with the properties of numbers and the solution of equations. One of the simplest examples of a nonarithmetical nature is elementary set algebra; we will also find it useful to use some of the language of sets throughout this book.†

A *set* is simply a collection of objects, finite or infinite, either exhibited explicitly or described in some definite fashion.

Example 1. The set $\{1, 4, 5\}$ consists of the integers 1, 4, and 5.

Example 2. The set of all names in the Seattle telephone directory as of November, 1964.

Example 3. The set of all the points in a plane.

Example 4. The set of all positive integers.

Often the so-called set-builder notation is used. Thus $\{x|x \text{ is a number and } x + 3 = 5\}$ is read, "The set of all x such that x is a number and $x + 3 = 5$." Clearly $\{x|x \text{ is a number and } x + 3 = 5\} = \{2\}$. Often phrases such as "x is a number" are left out and simply implied by the context. Thus $\{x|x + 3 = 5\} = \{x|x \text{ is a number and } x + 3 = 5\}$. Similarly the set of all numbers greater than 3 may be written $\{x|x > 3\}$.

A set is called *finite* if it can be put into one-to-one $(1-1)$ correspondence with the set of positive integers $\{1, 2, \ldots, n\}$ for some positive integer n; that is, if we can match up each element of the given set with each element of the set $\{1, 2, \ldots, n\}$. Thus the sets in Examples 1 and 2 are finite but the sets in Examples 3 and 4 are not finite and are called *infinite*.

It is convenient to introduce the convention of the *null* set, denoted by $\varnothing$, with no elements. (Do not confuse $\varnothing$ with the set $\{0\}$, which has the one element, 0, in it. For example, $\{x|x + 3 = 3\} = \{0\}$ but $\{x|x > 0 \text{ and } x + 3 = 3\} = \varnothing$.)

† In fact, many of the newer texts on elementary and secondary mathematics make extensive use of the concept of a set.

2. Set Membership and Set Inclusion

If x is an element of a set A we write $x \in A$; if $x \in A$ implies $x \in B$, we write $A \subseteq B$; if $A \subseteq B$ and $B \subseteq A$, we write $A = B$. That is, $A = B$ if every element of A is an element of B and if every element of B is an element of A. In symbols: $A = B$ if $x \in A$ implies $x \in B$ and $x \in B$ implies $x \in A$.

If $A \subseteq B$ we will say that A is a *subset* of B. Note that, by our definition, $A \subseteq A$ and hence A is a subset of A. But if $A \subseteq B$ and $B \neq A$, we say that A is a *proper* subset of B and write $A \subset B$. By $A \supseteq B$ we mean $B \subseteq A$ and by $A \supset B$, we mean $B \subset A$.

Example 1. If $A = \{1, 2, 3\}$, $1 \in A$.

Example 2. $\{1\} \subset \{1, 2, 3\}$.

Example 3. $\{1\} \notin \{1, 2, 3\}$ ("$\notin$" means "is not an element of").

Example 4. $1 \nsubseteq \{1, 2, 3\}$ ("$\nsubseteq$" means "is not a subset of").

Example 5. $\{1, 2, 1\} = \{1, 2\}$.

Example 6. $\{1, 2\} \subseteq \{1, 2\}$ but $\{1, 2\} \not\subset \{1, 2\}$.

Exercises 1.2

1. Decide which of the following statements are true and which are false.

(a) $1 \in \{1, 2\}$
(b) $\{1\} \in \{1, 2\}$
(c) $1 \subset \{1, 2\}$
(d) $\{1\} \subset \{1, 2\}$
(e) $\{1, 2\} \in \{1, 2, 1, 2, 1\}$
(f) $\{1, 2\} \supseteq \{1, 2, 1, 2, 1\}$
(g) $a \in \{a, \{a\}\}$
(h) $\{1, 2\} = \{1, 2, 1, 2, 1\}$
(i) $\varnothing \in \{\varnothing, \{\varnothing\}\}$
(j) $\varnothing \subset \{\varnothing, \{\varnothing\}\}$
(k) $\{1\} \in \{\{1\}, \{2\}\}$
(l) $\{1\} \subset \{\{1\}, \{2\}\}$
(m) $\{1, 1, 2\} = \{1, 2\}$
(n) $\{1, 2\} \supset \{1, 1, 2\}$
(o) $\{x | x$ is a number and $x^2 - 5x + 6 = 0\} = \{2, 3\}$.
(p) $\{x | x$ is an integer and x is divisible by $3\} = \{y | y$ is an integer and y is divisible by $3\}$.

2. Prove that if A, B, and C are sets such that $A \subseteq B$, $B \subseteq C$, and $C \subseteq A$, then $B = C$.

3. Prove that, for all sets A, $\varnothing \subseteq A$.

4. Suppose that A, B, and C are sets. Prove that if $A \subset B$ and $B \subset C$, then $A \subset C$.

5. Prove that if $\{x, y\} = \{x, z\}$, then $y = z$.

***6.** Does there exist a set A such that $A \in A$?

3. Union and Intersection of Sets

If A and B are two sets we define their *union*, $A \cup B$, to be the set of elements which are either in A or in B (or in both). That is, $x \in A \cup B$ if and only if $x \in A$ or $x \in B$.

Example 1. $A = \{1, 2, 3\}$, $B = \{4, 5, 6\}$; $A \cup B = \{1, 2, 3, 4, 5, 6\}$.

Example 2. $A = \{1, 2, 3\}$, $B = \{2, 3, 4\}$; $A \cup B = \{1, 2, 3, 4\}$.

Example 3. $A = \{x | x \text{ is an integer and } x \geq 3\}$, $B = \{x | x \text{ is an integer and } x \geq 6\}$; $A \cup B = A$.

By the *intersection*, $A \cap B$, of two sets, A and B, we mean the set consisting of the elements that are in both A and B. That is, $x \in A \cap B$ if and only if $x \in A$ and $x \in B$.

Example 4. $A = \{1, 2, 3\}$, $B = \{2, 3, 4\}$; $A \cap B = \{2, 3\}$.

Example 5. $A = \{1, 2, 3\}$, $B = \{4, 5, 6\}$; $A \cap B = \varnothing$.

Example 6. $A = \{x | x \geq 2\}$, $B = \{x | x \leq 5\}$; $A \cap B = \{x | 2 \leq x \leq 5\}$.

We are now in a position to prove many simple theorems in set theory such as:

1. $A \cap B = B \cap A$; $A \cup B = B \cup A$ (commutative laws).

2. $A \cap A = A$; $A \cup A = A$ (idempotent laws).

3. $A \cap (B \cap C) = (A \cap B) \cap C$; $A \cup (B \cup C) = (A \cup B) \cup C$ (associative laws).

4. $A \cap (B \cup C) = (A \cap B) \cup (A \cap C)$; $A \cup (B \cap C) = (A \cup B) \cap (A \cup C)$ (distributive laws).

5. $A \cap B \subseteq A$; $A \cup B \supseteq A$.

We will prove here that $A \cap B \subseteq A$ and that $A \cap (B \cap C) = (A \cap B) \cap C$ leaving the rest as exercises for the student.

For the first of these we simply remark that if $x \in A \cap B$, x is in A *and* in B and hence certainly in A. Thus $x \in A \cap B$ implies $x \in A$ and therefore $A \cap B \subseteq A$. For the second equality we note that if $x \in A \cap (B \cap C)$ it is in A and also in $B \cap C$. But if $x \in B \cap C$ it is in B and also in C. Hence $x \in A \cap (B \cap C)$ if and only if it is in A, B, and C. But if it is in A, B, and C, it is in $A \cap B$ and in C, and thus in $(A \cap B) \cap C$. Thus $A \cap (B \cap C) \subseteq (A \cap B) \cap C$. An entirely similar argument shows that $(A \cap B) \cap C \subseteq A \cap (B \cap C)$ and hence we may conclude that $A \cap (B \cap C) = (A \cap B) \cap C$.

Exercises 1.3

1. If $A = \{a, b, c\}$, $B = \{a, d, e\}$, and $C = \{d, f\}$, find
(a) $A \cap B$ (b) $C \cap A$
(c) $A \cap (B \cap C)$ (d) $A \cup (B \cap C)$
(e) $A \cap (B \cup C)$ (f) $(A \cup B) \cap (A \cup C)$.

2. Give an example of three sets A, B, and C such that $A \cap B \neq \varnothing$, $B \cap C \neq \varnothing$, and $A \cap C \neq \varnothing$ but $(A \cap B) \cap C = \varnothing$.

3. Prove the statements 1, 2, 4, and the second part of 5 given above.

4. Prove that $A \cap \varnothing = \varnothing$ and $A \cup \varnothing = A$ for all sets A.

5. Prove that $A \cap B = A$ if and only if $A \subseteq B$.

6. Suppose $A \subseteq S$. Define $\bar{A} = \{x | x \in S \text{ and } x \notin A\}$. ($\bar{A}$ is called the *complement* of A relative to S.)
(a) Show that if $B \subseteq S$, $\overline{A \cup B} = \bar{A} \cap \bar{B}$ and $\overline{A \cap B} = \bar{A} \cup \bar{B}$;
(b) What is $\bar{\varnothing}$ and what is $\bar{S}$?
(c) What is $\bar{\bar{A}}$?

***7.** Suppose S is a set and $A \subseteq S$, $B \subseteq S$. For $\bar{A}$ and $\bar{B}$ defined as in problem 6 we define $A/B = \bar{A} \cap \bar{B}$. Prove that
(a) $A/A = \bar{A}$
(b) $(A/B)/(A/B) = A \cup B$
(c) $(A/A)/(B/B) = A \cap B$.

4. Ordered Pairs

The set $\{a, b\}$ is equal to the set $\{b, a\}$. For an *ordered pair*, (a, b), however, we wish to make the requirement that $(a, b) = (c, d)$ if and only if $a = c$ and $b = d$. Thus, for example, $\{1, 2\} = \{2, 1\}$ but $(1, 2) \neq (2, 1)$.

We can do this in an elegant fashion by utilizing the concept of a set and defining $(a, b) = \{\{a\}, \{a, b\}\}$. To show, then, that $(a, b) = (c, d)$ implies $a = c$ and $b = d$ we first prove that

(1) $\{x, y\} = \{x, z\}$ implies $y = z$.

For, if $\{x, y\} = \{x, z\}$, we have $y \in \{x, z\}$ so that either $y = x$ or $y = z$. But, also, $z \in \{x, y\}$ so that $z = x$ or $z = y$. Thus we have $y = z$ or $x = y = z$.

Now suppose that $(a, b) = \{\{a\}, \{a, b\}\} = (c, d) = \{\{c\}, \{c, d\}\}$. Then $\{a\} \in \{\{c\}, \{c, d\}\}$ and hence $\{a\} = \{c\}$ or $\{a\} = \{c, d\}$. In either case $c \in \{a\}$ and hence $c = a$.

With $c = a$ we have $\{\{a\}, \{a, b\}\} = \{\{a\}, \{a, d\}\}$. By (1) it follows that $\{a, b\} = \{a, d\}$ and, again by (1), that $b = d$. On the other hand, if $a = c$ and $b = d$ it is obvious that $(a, b) = (c, d)$.

This shows that if we define $(a, b) = \{\{a\}, \{a, b\}\}$, then $(a, b) = (c, d)$ if and only if $a = c$ and $b = d$.

If S and T are sets, then the *Cartesian product*, $S \times T$, of S and T is the set of all ordered pairs (s, t) with $s \in S$ and $t \in T$. That is, $S \times T = \{(s, t) | s \in S, t \in T\}$. For example, if S and T are both taken to be the set of real numbers, then $S \times T$ is the set of all ordered pairs (a, b) where a and b are real numbers. The elements of this set correspond to the points in the Cartesian plane in the way utilized in analytic geometry; hence the name Cartesian product.

As another example, let $S = \{1, 2\}$ and $T = \{3, 4, 5\}$. Then $S \times T = \{(1, 3), (1, 4), (1, 5), (2, 3), (2, 4), (2, 5)\}$ and $T \times S = \{(3, 1), (3, 2), (4, 1), (4, 2), (5, 1), (5, 2)\}$. Note that here, and in general, $S \times T \neq T \times S$ unless $S = T$.

5. Binary Operations

If we have given a set S such that every ordered pair, (x, y), of elements $x, y \in S$ determines uniquely another element, $x * y$ in S, we say that we have defined a *binary operation* on S. For example, if S is the set of positive integers we might define $x * y$ as $x + y$ and observe that, since the sum of two positive integers is a positive integer, $x * y \in S$. On the other hand, if S is the set of odd positive integers and $x, y \in S$, $x + y$ is even so that $x + y \notin S$. Hence in this case we have not defined a binary operation on S. But if we define $x * y$ as xy we have a binary operation on the set of odd positive integers since the product of two odd integers is again an odd integer.

Exercises 1.5

Determine whether or not the following operations are binary operations on the given set S.

1. S = set of even integers; $x * y = x + y$.
2. S = set of even integers; $x * y = xy$.
3. S = set of nonzero integers; $x * y = x + y$.
4. S = set of subsets of the set of integers; $x * y = x \cap y$.
5. S = set of subsets of the set of integers; $x * y = x \cup y$.
6. S = set of positive integers; $x * y = x - y$.
7. S = set of integers; $x * y = x \div y$.
8. $S = \{0, 1\}$; $x * y = x + y$.
9. $S = \{0, 1\}$; $x * y = xy$.
10. $S = \{1/2^n | n \text{ a positive integer}\}$; $x * y = x + y$.

REFERENCES FOR FURTHER READING

[H2], [H3], [K1],]K2], [S3].

2

The Natural Numbers

1. The Problem

In the game of "Twenty Questions," A says, "I am thinking of something," and B asks A, for example, "Is it a mineral?" If the answer is "Yes," B tries a further question to find out what kind of a mineral it is, while if the answer is "No," he may ask, "Is it a vegetable?" Eventually, with skill and luck, B may arrive at an answer to what A is thinking of within the limits of the twenty allowable questions.

If I now pose my own variety of "Twenty Questions" and say to you, "I am thinking of a set on which are defined two binary operations," and I am thinking of the natural numbers—the ordinary counting numbers of arithmetic—1, 2, 3, . . . ,—can you ask the right questions to discover the answer? As we shall see at the end of the chapter, there is more than one set of questions which might be asked; the one we give now is only one possibility.

Stated in a more formal fashion, we ask: How may the natural numbers be uniquely characterized? Since the natural numbers may be used (as we shall see) to develop the other number systems used in elementary algebra, this is a basic question.

2. Closure

We are given a set S with two binary operations defined on it. Now since we know that we are going to end up with the natural numbers and the ordinary operations of addition and multiplication, it is natural to use the ordinary symbols for these operations. To do so, however, at this stage of the game is to prejudge the question. So let us hedge by using $\oplus$ and $\otimes$ to designate the two operations with the understanding that, at the moment, $a \otimes b$ might turn out to mean, for example, $a \div b$ for a and b nonzero rational numbers or that $a \oplus b$ might mean $a \cup b$ for a and b sets. Since we are requiring that $\oplus$ and $\otimes$ be binary operations on S we may list as our first postulate the requirement of *closure* with respect to these operations.

Postulate 1. If $a \in S$ and $b \in S$, then $a \oplus b \in S$ and $a \otimes b \in S$.

Exercises 2.2

Determine whether or not the following sets with the given operations satisfy Postulate 1.

1. S = rational numbers; $a \oplus b = ab$, $a \otimes b = a + b$.
2. S = rational numbers; $a \oplus b = a - b$, $a \otimes b = b - a$.
3. S = set of subsets of a given set A; $a \oplus b = a \cap b$, $a \otimes b = a \cup b$.
4. S = set of odd integers; $a \oplus b = a + b$, $a \otimes b = ab$.
5. S = set of even integers; $a \oplus b = a + b$, $a \otimes b = ab$.
6. S = set of rational numbers; $a \oplus b = a - b$, $a \otimes b = a \div b$.
7. S = set of nonzero rational numbers; $a \oplus b = a - b$, $a \otimes b = a \div b$.
8. S = set of natural numbers; $a \oplus b = 2a + b$, $a \otimes b = 2ab$.
9. S = set of irrational numbers; $a \oplus b = a + b$, $a \otimes b = ab$.
10. $S = \{0\}$; $a \oplus b = a + b$; $a \otimes b = ab$.

3. The Commutative and Associative Laws

Postulate 2. If $a \in S$ and $b \in S$, then $a \oplus b = b \oplus a$ and $a \otimes b = b \otimes a$ (the commutative laws).

Postulate 3. If $a \in S$, $b \in S$, and $c \in S$, then $(a \oplus b) \oplus c = a \oplus (b \oplus c)$ and $(a \otimes b) \otimes c = a \otimes (b \otimes c)$ (the associative laws).

Now if we take S to be the set of integers and $a \oplus b = a - b$ we see that Postulate 1 is satisfied but Postulate 2 is not, since, for example, $2 - 3 \neq 3 - 2$. To return to our game for a moment (and we shall not do so again explicitly) suppose your first question was, "Is it true that $a \oplus b = b \oplus a$ and $a \otimes b = b \otimes a$?" If I answer truthfully, "Yes," you would know now some of the sets together with binary operations that I was *not* thinking of! For example, I could not be thinking of the set of integers with $a \oplus b = a - b$. On the other hand, since the integers with $a \oplus b = a + b$ and $a \otimes b = ab$ satisfy Postulates 1 and 2, I could, as far as you would know from your question, be thinking of the integers under ordinary addition and multiplication instead of the natural numbers. In other words, Postulates 1 and 2 certainly do not uniquely characterize the natural numbers.

Now let $S = \{a, b, c\}$. (Note that it is *not* necessary to say anything about the "meaning" of a, b, and c!) Define $\oplus$ by the table

$\oplus$	a	b	c
a	a	c	b
b	c	a	c
c	b	c	a

By this we mean that the table is used as an ordinary addition table would be used: $a \oplus a = a$, $a \oplus b = c$ and so on. Finally, let $x \otimes y = a$ for all $x, y \in S$. It is easy to see that Postulates 1 and 2 are satisfied by this system, but

$$(a \oplus b) \oplus c = c \oplus c = a$$

while

$$a \oplus (b \oplus c) = a \oplus c = b$$

Note, however, that the rational numbers under ordinary addition and multiplication still satisfy all of our postulates, so that these three postulates still do not uniquely characterize the natural numbers.

Before going on to the next postulate we remark that it is possible to give examples of systems in which Postulate 1 holds and the first half of Postulate 2 ($a \oplus b = b \oplus a$) but not the second ($a \otimes b = b \otimes a$). Similarly, we can give examples of systems in which Postulates 1 and 2 and the first half of Postulate 3 hold, but not the second. Or, we can give examples of systems where Postulates 1 and 3 hold, but not 2, and so on. A great multiplicity of examples, however, may be confusing at this stage and if the student will simply have faith that such examples exist, nothing will have been lost by omitting discussion of them at this time. (See also the exercises below.)

Exercises 2.3

1. For the following sets and operations determine whether or not the operations are commutative and/or associative (both for $\oplus$ and for $\otimes$).
(a) S = rational numbers; $a \oplus b = ab/2$, $a \otimes b = (a + b)/2$.
(b) S = rational numbers; $a \oplus b = b$, $a \otimes b = a$.
(c) S = set of subsets of a given set A; $a \oplus b = a \cap b$, $a \otimes b = a \cup b$.
(d) S = set of integers; $a \oplus b = a - b$, $a \otimes b = ab$.
(e) S = set of integers; $a \oplus b = 2a + b$, $a \otimes b = 2ab$.

2. Find a set S and operations $\oplus$ and $\otimes$ on S such that Postulate 1 is satisfied and
(a) $a \oplus b = b \oplus a$ but, for at least some $a, b \in S$, $a \otimes b \neq b \otimes a$.
(b) $a \otimes b = b \otimes a$ but, for at least some $a, b \in S$, $a \oplus b \neq b \oplus a$.
(c) Postulate 2 is satisfied and $(a \oplus b) \oplus c = a \oplus (b \oplus c)$ but, for at least some $a, b, c \in S$, $a \otimes (b \otimes c) \neq (a \otimes b) \otimes c$.
(d) Postulate 3 is satisfied but one or both parts of Postulate 2 do not hold.

4. The Distributive Law for Multiplication with Respect to Addition

Postulate 4. If $a \in S$, $b \in S$, and $c \in S$, then $a \otimes (b \oplus c) = (a \otimes b) \oplus (a \otimes c)$.

If we let S be the set of integers and define $a \oplus b = ab$, $a \otimes b = a + b$ we see immediately that Postulates 1 to 3 are satisfied. Postulate 4 is not satisfied, however, since

$$2 \otimes (3 \oplus 4) = 2 + (3 \times 4) = 2 + 12 = 14$$

but

$$(2 \otimes 3) \oplus (2 \otimes 4) = (2 + 3) \times (2 + 4) = 5 \times 6 = 30.$$

We note, however, that the rational numbers under ordinary addition and multiplication also satisfy Postulates 1 to 4 so these four postulates still do not uniquely characterize the natural numbers.

Exercises 2.4

1. Determine whether or not Postulate 4 is satisfied for the sets and operations given in problem 1 of Exercises 2.3.

2. Find a set S closed under the operations $\oplus$ and $\otimes$ such that
(a) Postulates 2 and 4 are satisfied but one or both parts of Postulate 3 do not hold.
(b) Postulates 3 and 4 are satisfied but one or both parts of Postulate 2 do not hold.

5. The Remaining Postulates

Postulate 5. (the cancellation laws) If $a \in S$, $b \in S$, and $c \in S$, then
(1) if $a \oplus c = b \oplus c$, then $a = b$;
(2) if $a \otimes c = b \otimes c$, then $a = b$.

Postulate 5 has the effect of eliminating the rational numbers with $a \otimes b = ab$ from our list of examples, since, for example, $2 \times 0 = 3 \times 0$ but $2 \neq 3$. On the other hand, the *positive* rational numbers under ordinary addition and multiplication still continue to satisfy Postulate 5, as well as Postulates 1 to 4, so that we must conclude that Postulates 1 to 5 still do not uniquely characterize the natural numbers.

Postulate 6. (the existence of a unit for multiplication) There exists a unique element in S, which we will denote by the symbol 1, such that for every $a \in S$, $1 \otimes a = a$.

The set of all whole numbers greater than 1 under the usual definition of addition and multiplication satisfies Postulates 1 to 5 but not Postulate 6.

On the other hand, the positive rational numbers under ordinary addition and multiplication still satisfy Postulate 6, as well as Postulates 1 to 5.

We have seen that the set of positive rational numbers under ordinary addition and multiplication has been difficult to eliminate as a "candidate" for the title of a set of natural numbers in that it continues to satisfy all of the various postulates that we have laid down. We now outlaw it and almost all others by requiring that the *principle of finite* (or *mathematical*) *induction* hold.

***Postulate** 7.* If R is a subset of S such that $1 \in R$ and whenever $k \in R$ then $k \oplus 1 \in R$, then $R = S$.

This postulate is clearly satisfied by the natural numbers under ordinary addition since if R is a subset of the set of whole numbers and $1 \in R$, we then have $1 \oplus 1 = 2 \in R$, $2 \oplus 1 = 3 \in R, \ldots$; $R =$ the set of all natural numbers. On the other hand, let S be the set of positive rational numbers under ordinary addition and let R be the set of rational numbers

$$\{1, 1\tfrac{1}{2}, 2, 2\tfrac{1}{2}, 3, 3\tfrac{1}{2}, \ldots, n + \tfrac{1}{2}, \ldots\}$$

Then $1 \in R$ and whenever $k \in R$, $k \oplus 1 \in R$ but, certainly, $R \neq S$.

We still are able, however, to exhibit a peculiar system which satisfies Postulates 1 to 7 and which is not the set of natural numbers under ordinary addition and multiplication. Such a system is the set S consisting of the single element "1" when we define

$$1 \oplus 1 = 1 \otimes 1 = 1$$

To check Postulates 1 to 6 for this system is then very trivial since in every such postulate we must take $a = b = c = 1$, and every possible computation yields the answer of 1. Furthermore the only nonempty subset of S is S itself so that Postulate 7 is trivially satisfied.

***Postulate** 8.* (existence of solution of equations) If $a \in S$ and $b \in S$ then one and only one of the following alternatives hold: (1) $a = b$; (2) there exists an $x \in S$ such that $a \oplus x = b$; or (3) there exists a $y \in S$ such that $a = b \oplus y$.

In our system S described in connection with Postulate 7 consisting of 1 alone we have, for $a = b = 1$, not only (1) holding but also both (2) and (3) (take $x = y = 1$).

Now while it is true that in order for a set S with two binary operations defined on it to satisfy Postulates 1 to 8, S must be[†] the set of natural numbers under ordinary addition and multiplication, we as yet have no proof of this fact. It is true that we may examine set after set with varying definitions of $\oplus$ and $\otimes$ and not find another which satisfies all of our postulates, but since the number of sets is infinite, we can never arrive at a conclusion by this method.

We shall not attempt a proof here[‡] but simply assume the result. Or, to put it another way, we may *define* the natural numbers and addition and multiplication on them as the (unique)[§] set with the two binary operations of $\oplus$ and $\otimes$ which satisfy Postulates 1 to 8.

Exercises 2.5

Check each of the eight postulates for the following systems.

1. S = rational numbers; $a \oplus b = a + b$, $a \otimes b = ab$.
2. S = positive rational numbers; $a \oplus b = a + b$, $a \otimes b = ab$.
3. S = integers; $a \oplus b = a + b$, $a \otimes b = ab$.
4. S = natural numbers; $a \oplus b = 2(a + b)$, $a \otimes b = ab$.
5. $S = \{0, 1\}$; $a \oplus b = a$, $a \otimes b = ab$.
6. $S = \{0\}$; $a \oplus b = a + b$, $a \otimes b = ab$.
7. S = set of subsets of a given set A $(\neq \emptyset)$; $a \oplus b = a \cup b$, $a \otimes b = a \cap b$.
8. $S = \{a + b\sqrt{2} \mid a \text{ and } b \text{ integers}\}$; $a \oplus b = a + b$, $a \otimes b = ab$.
9. S = natural numbers; $a \oplus b = ab$, $a \otimes b = a + b$.
10. S = set of even natural numbers; $a \oplus b = a + b$, $a \otimes b = ab$.

6. Isomorphism

For precision we had to use the word "isomorphism" in the previous section in stating that the natural numbers were the "unique" set satisfying Postulates 1 to 8. Otherwise, for example, someone might claim we had two such sets:

$$S = \{1, 2, 3, 4, \ldots\} \text{ and } S' = \{\text{I}, \text{II}, \text{III}, \text{IV}, \ldots\}$$

Clearly, however, S and S' differ only in the symbols used in designating the elements of the set. More specifically, by the usual rules in regard to Roman numerals versus Arabic numerals, there is a one-to-one (1–1) correspondence between the symbols used to describe S and the symbols used to describe S'. Thus, for example, $1 \leftrightarrow \text{I}$, $2 \leftrightarrow \text{II}$, $19 \leftrightarrow \text{XIX}$. (It is true that the Roman numeral scheme tends to break down for larger numbers, but we can easily

† Up to an "isomorphism." This term will be defined precisely in the next section.
‡ See, for example, [B4].
§ Again, up to an isomorphism.

imagine an extension of it.) Furthermore, this correspondence is "preserved under addition and multiplication." For example,

$$2 \leftrightarrow \text{II} \qquad\qquad 3 \leftrightarrow \text{III}$$

Then

$$2 + 3 = 5 \qquad\qquad \text{II} + \text{III} = \text{V}$$

$$2 \times 3 = 6 \qquad\qquad \text{II} \times \text{III} = \text{VI}$$

and $5 \leftrightarrow \text{V}$, $6 \leftrightarrow \text{VI}$.

Let us contrast this with the situation where our two sets are: $S = \{1, 2, 3, 4, \ldots, n, \ldots\}$ and $T = \{2, 4, 6, 8, \ldots, 2n, \ldots\}$. It is still possible to set up a 1–1 correspondence between the elements of the two sets; for example, $1 \leftrightarrow 2$, $2 \leftrightarrow 4$, $3 \leftrightarrow 6$ etc. Then $n \leftrightarrow 2n$, $m \leftrightarrow 2m$ and, if the correspondence is preserved under addition, we must have both $n + m \leftrightarrow 2(n + m)$ and $n + m \leftrightarrow 2n + 2m$. Since $2n + 2m = 2(n + m)$ we conclude that the correspondence is indeed preserved under addition. But, for example

$$2 \leftrightarrow 4, \qquad\qquad 3 \leftrightarrow 6$$

with

$$2 \times 3 = 6, \qquad\qquad 4 \times 6 = 24$$

and we would want $6 \leftrightarrow 24$ if the correspondence is to be preserved under multiplication. But, under our defined correspondence, $6 \leftrightarrow 12$ and not to 24.

Our defined correspondence is not the only one possible, of course, but if, as we stated, the natural numbers are the unique set—up to an isomorphism—satisfying Postulates 1 to 8 then we cannot possibly have such a correspondence since the set T with its operations does not satisfy Postulate 6. Actually, it is easy to show in a more direct fashion that no 1–1 correspondence exists that is preserved under multiplication. Thus, suppose that we have such a correspondence in which, in particular, $1 \leftrightarrow 2k$ for some natural number k. Then we must have $1 \cdot 1 \leftrightarrow (2k) \cdot (2k)$. But $1 \cdot 1 = 1$ and hence $1 \leftrightarrow 4k^2$ so that $2k = 4k^2$ which is impossible.

We now make a formal definition of isomorphism.

Definition 1. *Let S and S' be two sets. On S are defined the binary operations $+$, $\times$, ... while on S' are defined the binary operations $+'$, $\times'$, ... Then we say that S and S' are isomorphic (with respect to the operations defined on them) if*

(1) *There is a 1–1 correspondence between the elements of S and S'.*

(2) *If* $a \leftrightarrow a'$ *and* $b \leftrightarrow b'$ $(a, b \in S;\ a', b' \in S')$ *under this correspondence, then* $a + b \leftrightarrow a' +' b'$, $a \times b \leftrightarrow a' \times' b', \ldots$.

Since we must have a 1–1 correspondence it is clear that two finite sets with a different number of elements cannot possibly be isomorphic.

We will consider the topic of isomorphisms again in greater detail in Section 7.6.

Exercises 2.6

1. Let $S = \{2, 4, 6, \ldots, 2n, \ldots\}$ and $S' = \{3, 6, 9, \ldots, 3n, \ldots\}$. Show that S and S' are isomorphic with respect to addition.

2. If $S = S'$, an isomorphism between S and S' is called an *automorphism* of S. If $S = \{a + b\sqrt{2} | a \text{ and } b \text{ integers}\}$ show that the correspondence $a + b\sqrt{2} \leftrightarrow a - b\sqrt{2}$ is an automorphism of S with respect to addition and multiplication.

***3.** Suppose that S is a set on which there is defined a binary operation $+$. Furthermore, suppose that there exists an element $e \in S$ with the property that $x + e = x$ for all $x \in S$. Finally, suppose that $+'$ is a binary operation on S' and that S and S' are isomorphic with respect to these operations. Show that there is an element $e' \in S'$ such that $x' + e' = x'$ for all $x' \in S'$ and that $e \leftrightarrow e'$.

***4.** Let $S = \{a + b\sqrt{2} | a \text{ and } b \text{ integers}\}$ and $S' = \{a + b\sqrt{3} | a \text{ and } b \text{ integers}\}$. Prove that S and S' are isomorphic with respect to addition but not with respect to multiplication.

***5.** In problem 1, show that S and S are isomorphic with respect to multiplication.

7. The Principle of Finite Induction

Let us return now to the principle of finite induction as a tool for the proof of propositions involving natural numbers. For example, consider the statement:

If n is a natural number, then

$$1 + 2 + \ldots + n = \frac{n(n+1)}{2}$$

To prove this we let S be the set of natural numbers, n, for which

$$1 + 2 + \ldots + n = \frac{n(n+1)}{2}$$

That is,

$$S = \{n \mid n \text{ a natural number and } 1 + 2 + \ldots + n = \frac{n(n+1)}{2}\}$$

(Conceivably, $S = \varnothing$ or $\{1\}$ or $\{1, 2\}$ and so on.) We want to prove that $S = N$, the set of all natural numbers. By the principle of finite induction, if (1) $1 \in S$ and (2) $k \in S$ implies $k + 1 \in S$, then $S = N$.

Now $1 \in S$ since 1 is a natural number and

$$1 = \frac{1(1+1)}{2}$$

Furthermore, if $k \in S$, then

$$1 + 2 + \ldots + k = \frac{k(k+1)}{2}. \tag{1}$$

Hence, adding $k + 1$ to both sides of (1), we have

$$\begin{aligned} 1 + 2 + \ldots + k + (k+1) &= \frac{k(k+1)}{2} + k + 1 \\ &= \frac{k^2 + k}{2} + \frac{2k + 2}{2} \qquad (2) \\ &= \frac{k^2 + 3k + 2}{2} = \frac{(k+1)(k+2)}{2} = \frac{(k+1)[(k+1)+1]}{2}. \end{aligned}$$

But (2) is the statement that

$$1 + 2 + \ldots + n = \frac{n(n+1)}{2}$$

when $n = k + 1$; i.e., that $k + 1 \in S$. Thus $S = N$.

We have considered this simple example in some detail to emphasize the rationale behind the manipulations. It is, of course, important that we show that $1 \in S$ *and* that $k \in S$ implies $k + 1 \in S$. Thus consider the (false) statement:

If n is a natural number, then $n^2 = n$.

If we let $S = \{n \mid n \text{ a natural number and } n^2 = n\}$ we do have $1 \in S$ but, of course, we do not have $k \in S$ implies $k + 1 \in S$. On the other hand, consider the (false) statement:

If n is a natural number, then $n + 1 = n$.

If we let $S = \{n \mid n \text{ a natural number and } n + 1 = n\}$ and assume $k \in S$ we have $k + 1 = k$. Hence $(k + 1) + 1 = k + 1$ which is $n + 1 = n$ for $n = k + 1$. Thus if $k \in S$, then $k + 1 \in S$. But, of course, $1 \notin S$.

In Section 5.5 we will consider further the uses of mathematical induction and present alternative forms of the principle.

Exercises 2.7

1. Use the principle of finite induction to establish the following equalities:
(a) $1^2 + 2^2 + \ldots + n^2 = n(n+1)(2n+1)/6$
(b) $1 \cdot 2 + 2 \cdot 3 + \ldots + n(n+1) = n(n+1)(n+2)/3$
(c) $1^3 + 2^3 + \ldots + n^3 = [n(n+1)/2]^2$
(d) $\dfrac{1}{1 \cdot 2} + \dfrac{1}{2 \cdot 3} + \ldots + \dfrac{1}{n(n+1)} = \dfrac{n}{n+1}$
(e) $1^2 + 4^2 + 7^2 + \ldots + (3n-2)^2 = n(6n^2 - 3n - 1)/2$
(f) $1^6 - 2^6 + 3^6 + \ldots + (-1)^{n+1}n^6 = \dfrac{(-1)^{n+1}}{2}(n^6 + 3n^5 - 5n^3 + 3n)$
(g) $\dfrac{1}{1 \cdot 2 \cdot 3} + \dfrac{1}{2 \cdot 3 \cdot 4} + \dfrac{1}{3 \cdot 4 \cdot 5} + \ldots + \dfrac{1}{n(n+1)(n+2)} = \dfrac{n(n+3)}{4(n+1)(n+2)}$
(h) $1 + 2 + 2^2 + \ldots + 2^{n-1} = 2^n - 1$
(i) $3 + 3^4 + 3^5 + \ldots + 3^{2n-1} = \frac{3}{8}(9^n - 1)$
(j) $\left(1 + \dfrac{1}{1}\right)\left(1 + \dfrac{1}{2}\right)\left(1 + \dfrac{1}{3}\right) \ldots \left(1 + \dfrac{1}{n}\right) = n + 1.$

2. Define $a^1 = a$ and $a^{k+1} = aa^k$ for $k \neq 1$, $k \in N$, and $a \in N$. Prove by induction that (a) $1^n = 1$; (b) $a^m a^n = a^{m+n}$; (c) $(a^m)^n = a^{mn}$ for all m, $n \in N$.

8. The Peano Postulates

At the beginning of this chapter we commented that there is more than one way to characterize the natural numbers. A very neat way is to replace our eight postulates by the following (due to G. Peano, [1858–1932]).

Our undefined terms are natural number, successor, and 1.

Postulate 1. 1 is a natural number.

Postulate 2. For each natural number n there exists a unique natural number, n', called the successor of n.

Postulate 3. 1 is not the successor of any natural number.

Postulate 4. If $n' = m'$, then $m = n$.

Postulate 5. If S is a set of natural numbers such that (1) $1 \in S$ and (2) $n \in S$ implies $n' \in S$, then S is the set of all natural numbers.

If we interpret x' to mean $x + 1$, it is easy to show that our previous postulates imply the Peano postulates. We leave this as an exercise for the reader. It is also possible, with the appropriate definitions of addition and multiplication, to show that the previous postulates are implied by the Peano postulates. For details the reader is referred to [B4] or [L1].

Exercises 2.8

1. Prove that Peano's postulates for the natural numbers are implied by the set of eight postulates previously given.

***2.** Formulate a suitable definition of the sum, $m + n$, of two natural numbers m and n in terms of the successor.

***3.** Formulate a suitable definition of the product, mn, of two natural numbers m and n in terms of the successor.

PROBLEMS FOR FURTHER INVESTIGATION

1. Investigate the possibility of a set of postulates that would characterize the integers, $\{\ldots, -2, -1, 0, 1, 2, \ldots\}$ as our postulates characterized the natural numbers. [B4], [M1]

2. Investigate the possibility of a set of postulates for the rational numbers. [K3], [M1]

3. Investigate the possibility of a set of postulates for the real numbers. [K3], [M1]

4. Show that the Peano postulates imply the original set of postulates given for the natural numbers. [B4], [K3]

5. The principle of mathematical induction enables us to prove such equalities as $1 + 2 + \ldots + n = n(n + 1)/2$. Can you devise a method for conjecturing the answers to such sums (whose validity can then be established by induction)? [M4]

REFERENCES FOR FURTHER READING

[A3], [H3], [K2], [K3], [L1], [M1], [R1].

3

Equivalent Pairs of Natural Numbers

We now have available the natural numbers. If we consider such equations as $x + 2 = 6$ and $x + 4 = 7$, we see that they have solutions, 4 and 3 respectively, that are natural numbers and that these solutions are determined by the ordered pairs (6, 2) and (7, 4) respectively. On the other hand, the equation $x + 6 = 2$ has no solution in natural numbers. Nevertheless the concept of the pair (2, 6) is just as meaningful (or nonmeaningful!) as the concept of the pair (6, 2) and it may at least serve as the symbolic solution to the equation $x + 6 = 2$. In other words we can set up a 1–1 correspondence between ordered pairs of natural numbers (a, b) and equations $x + b = a$ by $(a, b) \leftrightarrow x + b = a$. We will call (a, b) the symbolic solution to $x + b = a$.

Now there is no reason to be mysterious about our goal. Eventually we hope to arrive at a collection of sets of pairs of natural numbers which will provide us with a system isomorphic to the ordinary integers of algebra: $\ldots, -4, -3, -2, -1, 0, 1, 2, 3, 4, \ldots$. Thus, since the equations $x + 1 = 2$, $x + 2 = 3$, $x + 3 = 4, \ldots$ all have the solution $x = 1$, we will want to have the integer 1 be represented by the set of ordered pairs $\{(2, 1), (3, 2), (4, 3), \ldots\}$. Similarly we will want to represent the integer -1 by the set of ordered pairs $\{(1, 2), (2, 3), (3, 4), \ldots\}$. But (and this is the important point) we wish to base our discussion firmly on the properties of the natural numbers as postulated in Chapter 2. We shall now proceed to this task. For convenience we will let $N = \{x|x \text{ a natural number}\}$.

1. Equivalence

Let us go back to our reason behind the introduction of ordered pairs of natural numbers and suppose that we are working with equations whose solutions are natural numbers. In particular, suppose that the equation $x + b = a$ has a solution in natural numbers that is the same as the solution in natural numbers of the equation $x + d = c$. More precisely, suppose that $a, b, c,$ and d are natural numbers, $A = \{x|x \in N \text{ and } x + b = a\} = B = \{x|x \in N \text{ and } x + d = c\}$, and $A \neq \varnothing$. We claim that $A = B$ if and only if $a + d = c + b$.

To prove this, suppose first that $s \in A$. Then $s + b = a$. If, also, $t \in A$, then $t + b = a$ and hence $t + b = s + b$. Then, by the cancellation law for addition of natural numbers, $t = s$. Hence $A = \{s\}$. Similarly there is a natural number u such that $B = \{u\}$. Thus if $A = B$, then $s = u$ and we have $s + d = c$ as well as $s + b = a$. But then $(s + d) + b = c + b$ and since $(s + d) + b = s + (d + b)$ by the associative law of addition for natural numbers, we have

$$s + (d + b) = c + b$$

Continuing, we have

$$s + (b + d) = c + b \quad \text{by the commutative law of addition in } N$$

$$(s + b) + d = c + b \quad \text{by the associative law of addition in } N$$

$$a + d = c + b \quad \text{since } s + b = a$$

Thus if $A = B$, it follows that $a + d = c + b$.

Now suppose $a + d = c + b$. We wish to prove that $A = B$. Since $A = \{s\}$ and $B = \{u\}$ we have $s + b = a$ and $u + d = c$. Thus $(s + b) + d = a + d = c + b$. Continuing, we have

$$\begin{aligned} s + (b + d) &= c + b \quad \text{by the associative law of addition in } N \\ s + (d + b) &= c + b \quad \text{by the commutative law of addition in } N \\ (s + d) + b &= c + b \quad \text{by the associative law of addition in } N \\ s + d &= c \qquad\quad \text{by the cancellation law of addition in } N \end{aligned}$$

Hence $s \in B$ and since $B = \{u\}$, $s = u$ and $A = B$.

We have shown that if the two equations $x + b = a$ and $x + d = c$ have solutions in natural numbers, they have the same solutions if and only if $a + d = c + b$ and we have used only the properties of natural numbers in arriving at this result. Since the symbolic solutions have been indicated as (a, b) and (c, d) respectively, we are led to make the following definition.

Definition 1. *$(a, b) \sim (c, d)$ (read "(a, b) is **equivalent** to (c, d)") if and only if $a + d = c + b$.*

Examples. $(2, 1) \sim (4, 3)$ since $2 + 3 = 4 + 1$; $(1, 4) \sim (5, 8)$ since $1 + 8 = 5 + 4$. (Compare $a - b = c - d$ if and only if $a + d = c + b$.)

Note that we are not stating that $(a, b) = (c, d)$ if and only if $a + d = c + b$. Indeed we have previously said that $(a, b) = (c, d)$ if and only if $a = c$ and $b = d$. Equality is restricted to mean identity. When we write

$a = b$ for two natural numbers a and b we mean that a and b are identical natural numbers. Thus, for example $1 + 2 = 3$ means that "$1 + 2$" and "3" name the same number. When, however, we define $(a, b) \sim (c, d)$ if and only if $a + d = c + b$ we are defining "$\sim$" in terms of "$=$" and we have no guarantee that "$\sim$" has any of the properties of "$=$". Thus, for example, if we were to define $(a, b) \sim (c, d)$ if and only if $a + b = c$ we would not even have $(2, 1) \sim (2, 1)$ since $2 + 1 \neq 2$.

We note that if $(a, b) = (c, d)$ then $(a, b) \sim (c, d)$. For if $(a, b) = (c, d)$ then $a = c$ and $b = d$ and hence $a + d = c + b$ becomes $a + d = a + d$. Furthermore, we may prove that "$\sim$" does indeed possess some of the most significant properties of "$=$".

Theorem 1.
(1) $(a, b) \sim (a, b)$ ($\sim$ is *reflexive*);
(2) If $(a, b) \sim (c, d)$, then $(c, d) \sim (a, b)$ ($\sim$ is *symmetric*);
(3) If $(a, b) \sim (c, d)$ and $(c, d) \sim (e, f)$, then $(a, b) \sim (e, f)$ ($\sim$ is *transitive*).

Proof.
(1) $(a, b) \sim (a, b)$ since $a + b = a + b$.
(2) If $(a, b) \sim (c, d)$, then $(c, d) \sim (a, b)$ since if $a + d = c + b$ we have $c + b = a + d$.
(3) If $(a, b) \sim (c, d)$ and $(c, d) \sim (e, f)$, then $a + d = c + b$ and $c + f = e + d$. Hence

$(a + d) + e = (c + b) + e$	
$a + (d + e) = (c + b) + e$	by the associative law of addition in N
$a + (e + d) = (c + b) + e$	by the commutative law of addition in N
$a + (c + f) = (c + b) + e$	since $e + d = c + f$
$a + (f + c) = (c + b) + e$	by the commutative law of addition in N
$(a + f) + c = c + (b + e)$	by the associative law of addition in N
$(a + f) + c = (e + b) + c$	by the commutative law of addition in N
$a + f = e + b$	by the cancellation law of addition in N

But if $a + f = e + b$, then $(a, b) \sim (e, f)$ as desired.

We have written this proof out in great detail with reasons for each step. From now on we will condense steps and frequently omit reasons. Until the student is thoroughly familiar with this type of proof he should supply the omitted steps and give a reason for each step.

Exercises 3.1

1. Let S be the set of all triangles in the plane. If $a, b \in S$ we write $a \cong b$ if and only if a and b are congruent triangles. Is "$\cong$" reflexive?; symmetric?; transitive?
2. Let S be the set of all integers. If $a, b \in S$ we write $a \sim b$, if and only if $a - b$ is divisible by 7. Is "$\sim$" reflexive?; symmetric?; transitive?

3. Let S be the set of all natural numbers. If a, $b \in S$ we write $a < b$ if and only if there exists a natural number n such that $a + n = b$. Is "$<$" reflexive?; symmetric?; transitive?

4. Let S be the set of all ordered pairs, (a, b), of natural numbers. If (a, b), $(c, d) \in S$ we write $(a, b) \sim (c, d)$ if and only if $a = d$ and $b = c$. Is "$\sim$" reflexive?; symmetric?; transitive?

5. Let S be the set of rational numbers. If $a, b \in S$ we write $a \sim b$ if and only if $a^2 = b^2$. Is "$\sim$" reflexive?; symmetric?;, transitive?

2. Definition of Addition and Multiplication

If, again, $x + b = a$ and $y + d = c$ are equations solvable in natural numbers where a, b, c, and d are natural numbers, we have $(x + b) + (y + d) = a + c$. Now we apply the associative and commutative laws for the addition of natural numbers and obtain

$$(x + y) + (b + d) = a + c$$

(The student should supply the details.) Thus, since (a, b) is the symbolic solution of $x + b = a$ and (c, d) the symbolic solution of $y + d = c$ we see that $(a + c, b + d)$ is the symbolic solution of $z + (b + d) = a + c$ where $z = x + y$. Hence it is natural to make the following definition.

***Definition* 2.** $(a, b) \oplus (c, d) = (a + c, b + d)$.

***Example*.** $(2, 5) \oplus (1, 6) = (3, 11)$. (*cf.* $(-3) + (-5) = -8$.)

The student should note two items: (1) To avoid confusion between the newly defined addition of pairs of natural numbers and our addition in N we have used the symbol $\oplus$ for the defined addition; (2) our use of "$=$" rather than "$\sim$" is correct here since we are simply stating that $(a, b) \oplus (c, d)$ is another symbol for $(a + c, b + d)$.

Similarly we have, from $x + b = a$ and $y + d = c$ for natural numbers $x, y, a, b, c,$ and d,

(1) $xy + dx + by + bd = ac$.

(Again, the student should supply reasons beginning with $(x + b)(y + d) = ac$.)† But from $y + d = c$ we have $by + bd = bc$ and hence (1) becomes

(2) $xy + dx + bc = ac$.

Then from (2) we obtain

(3) $xy + dx + bd + bc = ac + bd$.

† Note that the very writing of $xy + dx + by + bd$ implies the existence of the associative law for addition. We do not need to worry whether $xy + dx + by + bd$ means $(xy + dx) + (by + bd)$ or $xy + [dx + (by + bd)]$ etc. since, by the associative law, all of these various expressions are equal.

Now we observe that from $x + b = a$ we obtain $dx + db = da$ or $dx + bd = ad$ so that (3) may be rewritten as

(4) $xy + (ad + bc) = ac + bd.$

Thus since (a, b) is the symbolic solution of $x + b = a$ and (c, d) the symbolic solution of $y + d = c$, we see that $(ac + bd, ad + bc)$ is the symbolic solution for $z + (ad + bc) = ac + bd$ where $z = xy$. Hence it is natural to make the following definition.

Definition 3. $(a, b) \otimes (c, d) = (ac + bd, ad + bc)$.

Example. $(2, 5) \otimes (1, 6) = (2 + 30, 12 + 5) = (32, 17)$. (Compare $(-3) \times (-5) = 15(= 32 - 17)$ and the remarks following Definition 2.)

As well motivated as our definitions are, we have one problem. For natural numbers, if $a = c$ and $b = d$ we have $a + b = c + d$ and $ab = cd$ from the very meaning of equality as identity. Here, however, $\sim$ is not identity and it is conceivable that we might have $(a, b) \sim (c, d)$ and $(e, f) \sim (u, v)$ and yet have $(a, b) \oplus (e, f) \nsim (c, d) \oplus (u, v)$ or $(a, c) \otimes (e, f) \nsim (c, d) \otimes (u, v)$. ($\nsim$ is read "not equivalent to.") Surely this would be a most undesirable state of affairs. Our next theorem states that this unpleasant situation does not occur.

Theorem 2. If $(a, b) \sim (c, d)$ and $(e, f) \sim (u, v)$, then

(1) $(a, b) \oplus (e, f) \sim (c, d) \oplus (u, v)$

and

(2) $(a, b) \otimes (e, f) \sim (c, d) \otimes (u, v)$.

Proof. We prove part (2) only leaving (1) as an exercise for the student. We have given $(a, b) \sim (c, d)$ and $(e, f) \sim (u, v)$ so that

(3) $a + d = c + b$ and $e + v = u + f$.

Now $(a, b) \otimes (e, f) = (ae + bf, af + be)$ and $(c, d) \otimes (u, v) = (cu + dv, cv + du)$. Thus (2) becomes $(ae + bf, af + be) \sim (cu + dv, cv + du)$ so that, by definition of $\sim$, (2) is equivalent to

(4) $(ae + bf) + (cv + du) = (cu + dv) + (af + be)$.

That is, to establish (2) we need to derive (4) from (3). We proceed.

From (3) we obtain, by the distributive law in N, $ae + de = ce + be$ and $ce + cv = cu + cf$, which then yield, by addition,

(5) $(ae + de) + (ce + cv) = (ce + be) + (cu + cf)$.

By a liberal use of the associative and commutative laws for addition in N we transform (5) into

(6) $(ae + de + cv) + ce = (be + cf + cu) + ce$.

Now we apply the cancellation law for addition in N to obtain

(7) $ae + de + cv = be + cf + cu$

which at least bears a faint resemblance to the desired (4). We continue. Again from (3) we obtain

(8) $cf + bf = af + df$ and $du + df = de + dv$.

Combining (7) and (8) by addition we have

$$(9) \quad (ae + de + cv) + (cf + bf) + (du + df) = (be + cf + cu) + (af + df) + (de + dv)$$

We apply the associative and commutative laws for addition in N once again to obtain, from (9),

$$[(ae + bf) + (cv + du)] + (de + cf + df) = [(cu + dv) + (af + be)] + (de + cf + df)$$

from which we obtain the desired (4) by the cancellation law.

Exercises 3.2

1. Use Definitions 2 and 3 to perform the following indicated operations:
(a) $(2, 1) \oplus (4, 6)$; $(2, 1) \otimes (4, 6)$
(b) $(3, 4) \oplus (1, 5)$; $(3, 4) \otimes (1, 5)$
(c) $(1, 2) \oplus (1, 3)$; $(1, 2) \otimes (1, 3)$
(d) $(5, 2) \oplus (6, 4)$; $(5, 2) \otimes (6, 4)$.

2. Verify Theorem 2 for the following cases:
(a) $(2, 1) \sim (4, 3)$; $(4, 6) \sim (1, 3)$
(b) $(3, 4) \sim (1, 2)$; $(1, 5) \sim (3, 7)$
(c) $(1, 2) \sim (4, 5)$; $(1, 3) \sim (2, 4)$
(d) $(5, 2) \sim (4, 1)$; $(6, 4) \sim (5, 3)$

3. Prove part (1) of Theorem 2.

3. Properties of Addition and Multiplication

Theorem 3. $(a, b) \oplus (c, d) = (c, d) \oplus (a, b)$

Proof.

$(a, b) \oplus (c, d) = (a + c, b + d)$	Definition 2
$(a + c, b + d) = (c + a, d + b)$	Commutative property of addition in N
$(c + a, d + b) = (c, d) \oplus (a, b)$	Definition 2

Theorem 4. $[(a, b) \oplus (c, d)] \oplus (e, f) = (a, b) \oplus [(c, d) \oplus (e, f)]$.

Proof.

$[(a, b) \oplus (c, d)] \oplus (e, f)$	$= (a + c, b + d) \oplus (e, f)$	Definition 2
$(a + c, b + d) \oplus (e, f)$	$= ([a + c] + e, [b + d] + f)$	Definition 2
$([a + c] + e, [b + d] + f)$	$= (a + [c + e], b + [d + f])$	Associative property of addition in N
$(a + [c + e], b + [d + f])$	$= (a, b) \oplus (c + e, d + f)$	Definition 2
$(a, b) \oplus (c + e, d + f)$	$= (a, b) \oplus [(c, d) \oplus (e, f)$	Definition 2

Theorem 5. $(a, b) \otimes (c, d) = (c, d) \otimes (a, b)$.

Proof.

$(a, b) \otimes (c, d)$	$= (ac + bd, ad + bc)$	Definition 3
$(ac + bd, ad + bc)$	$= (ca + db, da + cb)$	Commutative property of multiplication in N
$(ca + db, da + cb)$	$= (ca + db, cb + da)$.	Commutative property of addition in N
$(ca + db, cb + da)$	$= (c, d) \otimes (a, b)$	Definition 3

We thus see that the commutative laws of addition and multiplication and the associative law of addition for these pairs of natural numbers may be *proved* on the basis of our definitions and the postulates for the natural numbers.

The proofs for the associative law of multiplication and the distributive law for multiplication with respect to addition are entirely similar. But since they reflect no new principles and involve somewhat tedious computations we leave them as possible exercises for the student. We continue by considering whether or not there are theorems for the integers similar to the remaining postulates for the natural numbers.

Postulate 5 (the cancellation laws for addition and multiplication in N) needs some modification here. We give this as our next theorem.

Theorem 6. (1) If $(a, b) \oplus (c, d) \sim (e, f) \oplus (c, d)$, then $(a, b) \sim (e, f)$. (2) If $(a, b) \otimes (c, d) \sim (e, f) \otimes (c, d)$ and $c \neq d$, then $(a, b) \sim (e, f)$.

Proof of (1) Since $(a, b) \oplus (c, d) = (a + c, b + d)$ and $(e, f) \oplus (c, d) = (e + c, f + d)$ we see that our hypothesis implies

$$(a + c, b + d) \sim (e + c, f + d)$$

But then, by Definition 1,

$$(a + c) + (f + d) = (e + c) + (b + d)$$

We now use the associative and commutative properties for the natural numbers to obtain

$$(a + f) + (c + d) = (e + b) + (c + d)$$

from which, by the cancellation law of addition for natural numbers, we obtain

$$a + f = e + b$$

and hence

$$(a, b) \sim (e, f)$$

Proof of (2). Since $(a, b) \otimes (c, d) = (ac + bd, ad + bc)$ and $(e, f) \otimes (c, d) = (ec + fd, ed + fc)$ we see that our hypothesis implies that

$$(ac + bd, ad + bc) \sim (ec + fd, ed + fc)$$

But then

$$(ac + bd) + (ed + fc) = (ec + fd) + (ad + bc)$$

We now use the associative, commutative, and distributive properties for the natural numbers to obtain

$$(a + f)c + d(b + e) = (b + e)c + d(a + f)$$

Now if $c = d$ this reduces to an identity from which no conclusion can be drawn. But if $c \neq d$ (and this is part of our hypothesis), we have, by Postulate 8 for the natural numbers, that there exists a natural number x such that $c = d + x$ or there exists a natural number y such that $c + y = d$. Suppose that the first alternative holds. Then we have

$$(a + f)(d + x) + d(b + e) = (b + e)(d + x) + d(a + f)$$

Again using the appropriate properties of natural numbers we now obtain

$$(ad + ax + fd + fx) + (db + de) = (bd + bx + ed + ex) + (da + df)$$

or

$$(ax + fx) + (ad + fd + db + de) = (ex + bx) + (ad + fd + db + de)$$

Applying the cancellation law of addition for natural numbers we have

$$ax + fx = ex + bx$$

or

$$(a + f)x = (e + b)x$$

Now, finally, we apply the cancellation law of multiplication for natural numbers to obtain $a + f = e + b$ and, hence, $(a, b) \sim (e, f)$.

The case when there exists a natural number y such that $c + y = d$ presents no new difficulties and we leave this case to the student. (Note that the case of $c = d$ corresponds to $x \cdot 0 = y \cdot 0$ even when $x \neq y$.)

Using only single ordered pairs we have come about as far as we can in our efforts to develop the integers from the natural numbers. Thus if we examine our set of pairs of natural numbers for an analogy to Postulate 6 we find that (2, 1), (3, 2), (4, 3), and, in general, $(n + 1, n)$ all serve as unit elements for multiplication in the sense that, for example, $(2, 1) \otimes (a, b) = (2a + b, 2b + a) \sim (a, b)$.

We turn now, in the next chapter, to a consideration of equivalence classes and will construct the integers as classes of equivalent pairs of natural numbers.

Exercises 3.3

1. Complete the proof of part 2 of Theorem 6.
2. Prove the associative law of multiplication for these pairs of natural numbers.

3. Prove the distributive law for multiplication with respect to addition for these pairs of natural numbers.

***4.** Prove that if $(x, y) \otimes (a, b) \sim (a, b)$ for all pairs (a, b), then $(x, y) \sim (2, 1)$.

PROBLEMS FOR FURTHER INVESTIGATION

1. Suppose that S is again the set of all ordered pairs of natural numbers, (a, b), where now $(a, b) \sim (c, d)$ if and only if $a = c$ and $b = d$, $(a, b) \oplus (c, d) = (a + c, b + d)$, and $(a, b) \otimes (c, d) = (ac, bd)$. Provide an analogous treatment for S and its operations to the treatment just given.

2. Suppose that Mr. X dislikes working with "signed" numbers. He replaces all non-negative integers, a, by $2a + 1$ (e.g. 0 by 1, 1 by 3, 2 by 5 etc.) and all negative integers, $-a$, by $2a$ (e.g. -1 by 2, -2 by 4, -3 by 6 etc.). He then replaces the computations in the column on the left below with the computations in the column on the right.

$2 + 3 = 5$	$5 \oplus 7 = 5 + 7 - 1 = 11$
$(-2) + (-3) = -5$	$4 \oplus 6 = 4 + 6 = 10$
$(-2) + 3 = 1$	$4 \oplus 7 = 7 + 4 = 7 - 4 = 3$
$(-3) + 2 = -1$	$6 \oplus 5 = (6 - 5) + 1 = 2$

We note that, in Mr. X's scheme of things, the first answer of 5 in the left hand column does indeed correspond to his answer of 11 $[= (2 \times 5) + 1]$. Similarly, -5 corresponds to 10 $(= 2 \times 5)$, 1 to 3, and -1 to 2.

Can you describe in general his scheme for addition of "signed" numbers? Can you invent a similar scheme for multiplication? [D4].

REFERENCES FOR FURTHER READING

[H3], [L1], [M1], [R1], [T2].

4

Equivalence Classes and The Integers

1. Relations

The subject of *relations* in mathematics is a very large one. Our major concern here will be with the special class of relations, known as *equivalence relations*, which may be considered as a generalization of the idea of equality.

For example, equality is a relation on the natural numbers: If a and b are natural numbers, then either $a = b$ or $a \neq b$. Another relation on N is that of being less than: If a and b are natural numbers, then $a < b$ or $a \nless b$. Suppose now that we consider the set of all triangles in a plane and the relationship of similarity: If A and B are two triangles, then $A \sim B$ or $A \nsim B$.

To generalize these examples we let S be any set and, for $x, y \in S$, we write $x\,R\,y$ to mean that x is in the relation R to y and $x \not R\, y$ to mean that x is not in the relation R to y. Thus if $S = N$ and R is the relationship of less than, we have $2\,R\,3$ but $3 \not R\, 2$.

Definition 1. *R is said to be a **relation** on a set S if $x, y \in S$ implies either $x\,R\,y$ or $x \not R\, y$, but not both.*

It should be evident that the concept of a relation is a very broad one indeed. All that we require is that, given two elements x and $y \in S$, we be able to determine whether or not $x\,R\,y$.

Definition 2. *A relation R on a set S is said to be an **equivalence relation** if*

(1) *$a\,R\,a$ for all $a \in S$* (***reflexive*** *property*)
(2) *If $a\,R\,b$, then $b\,R\,a$* (***symmetric*** *property*)
(3) *If $a\,R\,b$ and $b\,R\,c$, then $a\,R\,c$* (***transitive*** *property*)

The relation of identity in any set is clearly an equivalence relation and hence, in particular, equality of natural numbers is an equivalence relation. Furthermore, the relation $\sim$ as defined in Chapter 3 for pairs of natural numbers is an equivalence relation (Theorem 1, Section 3.1), as is similarity for triangles in a plane. On the other hand, the relationship of less than on the natural numbers is not an equivalence relation for, while we do have the transitive property ($a < b$ and $b < c$ implies $a < c$), we do not have the reflexive ($a < a$) or symmetric ($a < b$ implies $b < a$) properties.

Exercises 4.1

1. Determine, for the following sets and relations, whether or not the relation is (i) reflexive, (ii) symmetric, and (iii) transitive.
   (a) $S = N$; $a\,R\,b$ if and only if a divides b.
   (b) $S =$ integers; $a\,R\,b$ if and only if $a - b$ is divisible by 3.
   (c) $S =$ integers; $a\,R\,b$ if and only if $a \leq |b|$.
   (d) $S =$ integers; $a\,R\,b$ if and only if $a^2 = b^2$.
   (e) $S =$ lines in a plane; $a\,R\,b$ if and only if a is perpendicular to b.
   (f) $S =$ lines in a plane; $a\,R\,b$ if and only if a is parallel to b.
   (g) $S =$ set of all people; $a\,R\,b$ if and only if a is a sibling of b.
   (h) $S =$ set of all people; $a\,R\,b$ if and only if a is a spouse of b.

2. A relation is called *circular* if $a\,R\,b$ and $b\,R\,c$ implies $c\,R\,a$. Show that a relation is reflexive and circular if and only if it is reflexive, symmetric, and transitive.

3. Let $S = \{(a, b) | a, b \in N\}$ and define $(a, b) \sim (c, d)$ if and only if $ad = bc$. Prove that $\sim$ is an equivalence relation on S.

4. Find sets S and relations on them such that the relation is
   (a) Reflexive but not symmetric or transitive
   (b) Reflexive and symmetric but not transitive
   (c) Reflexive and transitive but not symmetric
   (d) Symmetric but not reflexive or transitive
   (e) Symmetric and transitive but not reflexive
   (f) Transitive but not reflexive or symmetric.

5. A relation R on a set S is said to *partially order* the set S if for all a, b, c in S (i) $a\,R\,a$; (ii) if $a\,R\,b$ and $b\,R\,a$, then $a = b$; (iii) if $a\,R\,b$ and $b\,R\,c$, then $a\,R\,c$.
   (a) Give an example of a partially ordered set;
   (b) If we define a relation R' on S by $a\,R'\,b$ if and only if $b\,R\,a$, show that R' partially orders S if and only if R does.

6. What is wrong with the following argument showing that a relation which is both symmetric and transitive is necessarily reflexive? $a\,R\,b$ implies $b\,R\,a$ by symmetry; by transitivity, $a\,R\,b$ and $b\,R\,a$ implies $a\,R\,a$.

7. Let $S = \{n | n \in N \text{ and } n > 1\}$. If a, b $\in S$ define $a \sim b$ to mean that a and b have the same number of positive prime factors (distinct or identical). Show that $\sim$ is an equivalence relation defined on S.

2. Partitioning of a Set

Definition 3. *A set S is said to be* ***partitioned*** *by the sets $A, B, C, \ldots$ (the* ***partitioning classes****) if*
(1) $A \cup B \cup C \cup \ldots = S$
(2) *For any two partitioning classes X and Y we have $X = Y$ or $X \cap Y = \varnothing$*

Example 1. Let $S = \{1, 2, 3, 4, 5\}$ and $A = \{1, 2\}$, $B = \{3, 5\}$, $C = \{4\}$. Then S is partitioned by A, B, and C.

Example 2. Let $S = N$ and $A = \{x | x \in N$ and x is an even number$\}$, $B = \{x | x \in N$ and x is an odd number$\}$. Then S is partitioned by A and B.

The number of partitioning classes need not be finite. Thus if $S = N$ we have the infinite number of sets $\{1\}, \{2\}, \ldots, \{k\}, \ldots$ as a (trivial) set of partitioning classes. Other examples of the use of an infinite number of partitioning classes will be given later.

The general way of arriving at a partition of a set S is by the use of an equivalence relation on S as described in the following basic theorem.

Theorem 1. Let S be a set and R an equivalence relation on S. For any $a \in S$ let $P_a = \{x | x \in S$ and $x \, R \, a\}$. Then the set of all such P_a is a partition of S. Furthermore, if $P_1, P_2, \ldots,$ is a partition of S, then there exists a unique equivalence relation R on S such that $P_i = \{x | x \in S$ and $x \, R \, a_i\}$ for some $a_i \in S$ $(i = 1, 2, \ldots)$.

Proof. I. Suppose that R is an equivalence relation on S.

(1) We observe first that if $a \in S$, then $a \in P_a$ since $a \, R \, a$. Thus if we consider all possible sets P_y for any $y \in S$ we will obtain all of the elements of S at least once. That is, condition (1) of Definition 3 is satisfied:

$$\bigcup_{y \in S} P_y = P_a \cup P_b \cup P_c \cup \ldots = S$$

(2) Suppose that we have two partitioning classes P_a and P_b such that that $P_a \cap P_b \neq \varnothing$. We wish to show that $P_a = P_b$ (condition (2) of Definition 3). Since $P_a \cap P_b \neq \varnothing$ there exists an $x \in S$ such that $x \in P_a$ and $x \in P_b$. Now consider any element $y \in P_a$. Then $y \, R \, a$. Now $x \, R \, a$ and $x \, R \, b$ because $x \in P_a$ and $x \in P_b$. But $x \, R \, a$ implies $a \, R \, x$ (symmetric property) and $y \, R \, a$ and $a \, R \, x$ implies $y \, R \, x$ (transitive property). But also, $x \, R \, b$, and again applying the transitive property (now to $y \, R \, x$ and $x \, R \, b$) we obtain $y \, R \, b$, so that $y \in P_b$. Thus $y \in P_a$ implies $y \in P_b$ and $P_a \subseteq P_b$. Similarly, we obtain $P_b \subseteq P_a$ and hence $P_a = P_b$ as desired.

II. Now let $P_1, P_2, \ldots$ be a partition of S and define for $x, y \in S$, $x \, R \, y$ if and only if x and y are in the same partitioning class P_i. It is easy to verify that R is then an equivalence relation on S and this is left as an exercise for the student. If R' were another equivalence relation yielding the same partition $P_1, P_2, \ldots$, our definition of the partitioning classes as the sets $\{x | x \in S$ and $x\ R'\ a\}$ shows that $x\ R'\ y$ if and only if $x\ R\ y$ since $x\ R\ y$ if and only if x and y are in the same partitioning class. That is, R and R' are the same.

Example 3. Let $S = N$ and let $a\ R\ b$ if and only if a and b are both odd or both even. Thus $2\ R\ 6$ and $3\ R\ 5$ but $2\ \not R\ 3$ and $3\ \not R\ 6$. Since all natural numbers are either even or odd we have two partitioning classes,

$$P_1 = \{x | x \in N \text{ and } x\ R\ 1\} \text{ and } P_2 = \{x | x \in N \text{ and } x\ R\ 2\}$$

Example 4. Let $S = N$ and let $a\ R\ b$ if and only if $a = b$. Then $P_n = \{x | x \in N$ and $x\ R\ n\} = \{x | x \in N$ and $x = n\} = \{n\}$ and we have an infinite number of partitioning classes, each consisting of a single natural number.

Example 5. Let S be the set of all triangles in a plane and let R be the relation of similarity. Then each partitioning class of S is a set of similar triangles. There are an infinite number of partitioning classes and each partitioning class contains an infinite number of elements.

Example 6. Let $S = N$ with the partition $P_1 = \{x | x \in S$ and $x = 3n$ for $n \in N\}$, $P_2 = \{x | x \in S$ and $x = 3n + 1$ for $n \in N\}$, and $P_3 = \{x | x \in S$ and $x = 3n + 2$ for $n \in N\}$. The relation $x\ R\ y$ if and only if x and y are in the same P_i $(i = 1, 2, 3)$ is an equivalence relation on S. It may be restated as $x\ R\ y$ if and only if x and y have the same remainder when divided by 3.

Exercises 4.2

1. Describe the partitioning classes of the set of integers under the equivalence relation $a\ R\ b$ if and only if $a - b$ is divisible by 5.

2. Consider the set S and the equivalence relation on it as given in problem 3 of Exercises 4.1. Describe the partitioning classes under this equivalence relation.

3. Consider a set S and a relation R on it. Define, as before, $P_a = \{x | x \in S$ and $x\ R\ a\}$ for any $a \in S$. Show, by examples, that the set of all such P_a is not a partition of S if (a) R is reflexive and transitive but not symmetric; (b) R is symmetric and transitive but not reflexive; (c) R is reflexive and symmetric but not transitive.

4. Give the details of the verification that, in the proof of the second part of Theorem 1, R is an equivalence relation on S.

3. Constructing the Integers

In Chapter 3 we defined $(a, b) \sim (c, d)$ if and only if $a + d = c + b$ and showed (Theorem 3.1) that $\sim$ was an equivalence relation. Hence, by Theorem 4.1, $\sim$ defines a partition of the set of ordered pairs of natural numbers. Let us list some of these partitioning classes and attach the common names of these classes:

$$-2 = \{(1, 3), (2, 4), \ldots, (a, a + 2), \ldots\}$$

$$-1 = \{(1, 2), (2, 3), \ldots, (a, a + 1), \ldots\}$$

$$0 = \{(1, 1), (2, 2), \ldots, (a, a), \ldots\}$$

$$+1 = \{(2, 1), (3, 2), \ldots, (a + 1, a), \ldots\}$$

$$+2 = \{(3, 1), (4, 2), \ldots, (a + 2, a), \ldots\}$$

In general, for k a natural number, we define

$$+k = \{(a, b) | (a, b) \sim (k + 1, 1)\}$$

$$-k = \{(a, b) | (a, b) \sim (1, k + 1)\}$$

$$0 = \{(a, b) | (a, b) \sim (1, 1)\}$$

Note that we have not said that $k = +k$; k is a natural number and $+k$ is a set of pairs of natural numbers so we could not possibly have $k = +k$. Later on, however, we will show that the set $\{+k | k \text{ a natural number}\}$ is isomorphic to the set $\{k | k \text{ a natural number}\}$ under addition and multiplication. Thus we are now considering an integer as a *set* of ordered pairs.

Definition 4. *An* ***integer****, $I\langle a, b\rangle$, where a and b are natural numbers is defined by*

$$I\langle a, b\rangle = \{(x, y) | x, y \in N \textit{ and } (x, y) \sim (a, b)\}$$

Note that $(a, b) \in I\langle a, b\rangle$ since $(a, b) \sim (a, b)$.

Theorem 2. $I\langle a, b\rangle = I\langle c, d\rangle$ if and only if $a + d = c + b$.

Proof. If $I\langle a, b\rangle = I\langle c, d\rangle$ then, since $(a, b) \in I\langle a, b\rangle$, we have $(a, b) \in I\langle c, d\rangle$. Hence $(a, b) \sim (c, d)$ and thus $a + d = c + b$.

On the other hand, if $a + d = c + b$, then $(a, b) \sim (c, d)$. If $(x, y) \in I\langle a, b\rangle$, $(x, y) \sim (a, b)$. By the transitive property of $\sim$ we have $(x, y) \sim (c, d)$ and hence $(x, y) \in I\langle c, d\rangle$. Thus $I\langle a, b\rangle \subseteq I\langle c, d\rangle$. Similarly, we obtain $I\langle c, d\rangle \subseteq I\langle a, b\rangle$ and hence $I\langle a, b\rangle = I\langle c, d\rangle$.

Note that we are using the sign of equality in the proper sense of identity; the two sets $I\langle a, b\rangle$ and $I\langle c, d\rangle$ are identical if $a + d = c + b$.

Example 1. $I\langle 5, 4\rangle = I\langle 2, 1\rangle$; $I\langle 7, 13\rangle = I\langle 1, 7\rangle$.

Definition 5. $$I\langle a, b\rangle + I\langle c, d\rangle = I\langle a + c, b + d\rangle.$$

Definition 6. $$I\langle a, b\rangle \times I\langle c, d\rangle = I\langle ac + bd, ad + bc\rangle.$$

Here, of course, we should use, say, $\oplus$ and $\otimes$ as in Chapter 3 to distinguish the symbols used for the addition and multiplication of integers from the symbols used for addition and multiplication of natural numbers. At this point, however, no confusion is likely to result from the two distinct uses of the same symbols.

Example 2. $-2 = I\langle 1, 3\rangle$, $+3 = I\langle 4, 1\rangle$. Then
$(-2) + (+3) = I\langle 1, 3\rangle + I\langle 4, 1\rangle = I\langle 5, 4\rangle = I\langle 2, 1\rangle = +1$.
Also $(-2) \times (+3) = I\langle 1, 3\rangle \times I\langle 4, 1\rangle = I\langle 7, 13\rangle = I\langle 1, 7\rangle = -6$.

Before we can use Definitions 5 and 6 freely, however, we must prove the following theorem.

Theorem 3. If $I\langle a, b\rangle = I\langle a', b'\rangle$ and $I\langle c, d\rangle = I\langle c', d'\rangle$, then (1) $I\langle a, b\rangle + I\langle c, d\rangle = I\langle a', b'\rangle + I\langle c', d'\rangle$ and (2) $I\langle a, b\rangle \times I\langle c, d\rangle = I\langle a', b'\rangle \times I\langle c', d'\rangle$.

Since $I\langle a, b\rangle = I\langle a', b'\rangle$ if and only if $a + b' = a' + b$ [that is, if and only if $(a, b) \sim (a', b')$] it should be clear from this and Definitions 5 and 6 that Theorem 3 is simply a rephrasing of Theorem 3.2.

Exercises 4.3

1. Perform the indicated operations and express your answer in the form $I\langle 1, 1\rangle$, $I\langle k, 1\rangle$, or $I\langle 1, k\rangle$.
(a) $I\langle 2, 3\rangle + I\langle 4, 7\rangle$
(b) $I\langle 3, 1\rangle + I\langle 5, 8\rangle$
(c) $I\langle 1, 1\rangle + I\langle 5, 2\rangle$
(d) $I\langle 6, 2\rangle + I\langle 8, 3\rangle$
(e) $I\langle 2, 3\rangle \times I\langle 4, 7\rangle$
(f) $I\langle 3, 1\rangle \times I\langle 5, 8\rangle$
(g) $I\langle 1, 1\rangle \times I\langle 5, 2\rangle$
(h) $I\langle 6, 2\rangle \times I\langle 8, 3\rangle$

2. Verify Theorem 3 in the following cases.
(a) $I\langle 2, 3\rangle = I\langle 5, 6\rangle$, $I\langle 4, 7\rangle = I\langle 1, 4\rangle$
(b) $I\langle 2, 1\rangle = I\langle 5, 4\rangle$, $I\langle 5, 8\rangle = I\langle 2, 5\rangle$
(c) $I\langle 1, 1\rangle = I\langle 3, 3\rangle$, $I\langle 5, 2\rangle = I\langle 7, 4\rangle$
(d) $I\langle 6, 2\rangle = I\langle 8, 4\rangle$, $I\langle 8, 3\rangle = I\langle 6, 1\rangle$

4. Further Properties of the Integers

From Definitions 4 to 6, Theorem 3, and the results of Chapter 3, it is evident that the integers obey the commutative, associative, and distributive laws and that the cancellation laws hold as modified in Theorem 3.6. That is, if $I\langle a, b\rangle + I\langle c, d\rangle = I\langle e, f\rangle + I\langle c, d\rangle$, then $I\langle a, b\rangle = I\langle e, f\rangle$ and if $I\langle a, b\rangle \times I\langle c, d\rangle = I\langle e, f\rangle \times I\langle c, d\rangle$ and $c \neq d$, then $I\langle a, b\rangle = I\langle e, f\rangle$. We now consider some additional properties of the integers.

Theorem 4. There is a unique multiplicative identity, $+1$, for the integers.

Proof. We recall that $+1 = I\langle 2, 1\rangle$. Then $I\langle 2, 1\rangle \times I\langle a, b\rangle = I\langle 2a + b, a + 2b\rangle = I\langle a, b\rangle$. Furthermore, if $I\langle r, s\rangle \times I\langle a, b\rangle = I\langle a, b\rangle$ for all natural numbers a and b, we have $I\langle r, s\rangle \times I\langle a, b\rangle = I\langle 2, 1\rangle \times I\langle a, b\rangle$ for all natural numbers a and b. Taking $a \neq b$ we may use the cancellation law for multiplication to obtain $I\langle r, s\rangle = I\langle 2, 1\rangle$.

The student should compare the situation here very carefully with the situation in Chapter 3. There we pointed out that, for example, both (2, 1) and (3, 2) were multiplicative identities and, while $(2, 1) \sim (3, 2)$, $(2, 1) \neq (3, 2)$. Now we have $I\langle 2, 1\rangle = I\langle 3, 2\rangle = \{(2, 1), (3, 2), (4, 3), \ldots, (k + 1, k), \ldots\}$. Thus Postulate 6 for the natural numbers continues to hold for the integers.

Our next theorem is a modification of Postulate 8 for the natural numbers. In the proof of this theorem and elsewhere the following lemma is useful.

Lemma. $I\langle x + c, x + d\rangle = I\langle c, d\rangle$.

Proof. By Theorem 2, $I\langle x + c, x + d\rangle = I\langle c, d\rangle$ if and only if $(x + c) + d = c + (x + d)$. But

$$\begin{aligned}
(x + c) + d &= x + (c + d) && \text{Associative law of addition in } N\\
&= x + (d + c) && \text{Commutative law of addition in } N\\
&= (x + d) + c && \text{Associative law of addition in } N\\
&= c + (x + d) && \text{Commutative law of addition in } N
\end{aligned}$$

Theorem 5. If $I\langle a, b\rangle$ and $I\langle c, d\rangle$ are two integers, there exists a unique integer, $I\langle x, y\rangle$ such that $I\langle a, b\rangle + I\langle x, y\rangle = I\langle c, d\rangle$. We have $I\langle x, y\rangle = I\langle b + c, a + d\rangle$.

Proof.

$$\begin{aligned}
I\langle a, b\rangle + I\langle b + c, a + d\rangle &= I\langle a + (b + c), b + (a + d)\rangle\\
&\qquad \text{Definition of addition}\\
&= I\langle (a + b) + c, (b + a) + d\rangle\\
&\qquad \text{Associative law of addition in } N\\
&= I\langle (a + b) + c, (a + b) + d\rangle\\
&\qquad \text{Commutative law of addition in } N\\
&= I\langle c, d\rangle \quad \text{Lemma}
\end{aligned}$$

It remains to show that $I\langle x, y\rangle$ is unique. Now if $I\langle a, b\rangle + I\langle r, s\rangle = I\langle c, d\rangle$ we have $I\langle a, b\rangle + I\langle x, y\rangle = I\langle a, b\rangle + I\langle r, s\rangle$. Then $I\langle x, y\rangle + I\langle a, b\rangle = I\langle r, s\rangle + I\langle a, b\rangle$ and, by the cancellation law of addition for integers, we have $I\langle x, y\rangle = I\langle r, s\rangle$.

We have discussed, with respect to the integers, all of the eight postulates given for the natural numbers except Postulate 7, the principle of finite induction. That this is not a property of the integers is easily seen by considering the set $S = \{+k|k \text{ a natural number}\}$ that is a subset of the set I of all integers. Then $+1$, the multiplicative identity of I, is in S and if $+k \in S$, then $(+k) + (+1) = +(k + 1) \in S$. But $S \neq I$ since, for example, $0 \notin S$.

Exercises 4.4

1. Verify Theorem 5 by explicitly exhibiting an $I\langle x, y\rangle$ for the following cases:
   (a) $I\langle a, b\rangle = \langle 3, 4\rangle, I\langle c, d\rangle = \langle 3, 4\rangle$
   (b) $I\langle a, b\rangle = \langle 4, 2\rangle, I\langle c, d\rangle = \langle 6, 2\rangle$
   (c) $I\langle a, b\rangle = \langle 7, 3\rangle, I\langle c, d\rangle = \langle 2, 3\rangle$
   (d) $I\langle a, b\rangle = \langle 1, 4\rangle, I\langle c, d\rangle = \langle 6, 2\rangle$
   (e) $I\langle a, b\rangle = \langle 2, 5\rangle, I\langle c, d\rangle = \langle 5, 8\rangle$

2. Restate Theorems 3.3, 3.4, 3.5, and 3.6 using the notation $I\langle a, b\rangle$ in place of (a, b) etc.

5 The Natural Numbers as a Subset of the Integers

We have already remarked that the integers $+1, +2, +3, \ldots$ as defined in Section 3 are not the same as the natural numbers $1, 2, 3, \ldots$. We do, however, have

Theorem 6. The subset $\{+1, +2, \ldots, +k, \ldots\}$ of the integers $\{\ldots, -2, -1, 0, +1, +2, \ldots\}$ is isomorphic to the set of natural numbers $\{1, 2, 3, \ldots\}$ under the correspondence $+k \leftrightarrow k$ where k is a natural number and $+k = I\langle k + 1, 1\rangle$.

Proof. It is obvious that the correspondence is 1–1. Now suppose that $+k = I\langle k + 1, 1\rangle \leftrightarrow k$ and $+m = I\langle m + 1, 1\rangle \leftrightarrow m$. Then $(+k) + (+m) = I\langle k + 1, 1\rangle + I\langle m + 1, 1\rangle = I\langle k + m + 2, 2\rangle = I\langle (k + m) + 1, 1\rangle = +(k + m) \leftrightarrow k + m$ as desired. Similarly, $(+k) \times (+m) = I\langle k + 1, 1\rangle \times I\langle m + 1, 1\rangle = I\langle (k + 1)(m + 1) + 1, (k + 1) + (m + 1)\rangle = I\langle km + k + m + 2, k + m + 2\rangle = I\langle km + 1, 1\rangle = +(km) \leftrightarrow km$ as desired.

Exercises 4.5

1. Show that the correspondence $+k \leftrightarrow k$ of Theorem 6 is the only one possible if the correspondence is to be preserved under addition and multiplication. Is any other correspondence possible if we require preservation under addition only?

2. Prove that the subset $\{-1, -2, \ldots, -k, \ldots\}$ of the integers is isomorphic to the set of natural numbers under addition but not under multiplication.

***3.** In problem 1 is any other correspondence possible if we require preservation under multiplication only?

6. Zero and Subtraction

Theorem 7. Let $I\langle a, b\rangle$ be any integer. Then $0 + I\langle a, b\rangle = I\langle a, b\rangle + 0 = I\langle a, b\rangle$; that is, 0 is an additive identity. Furthermore, if $I\langle x, y\rangle + I\langle a, b\rangle = I\langle a, b\rangle$, then $I\langle x, y\rangle = 0$.

Proof. Recall that $0 = I\langle 1, 1\rangle$. Thus

$$0 + I\langle a, b\rangle = I\langle 1, 1\rangle + I\langle a, b\rangle = I\langle a + 1, b + 1\rangle = I\langle a, b\rangle,$$

and since $I\langle a, b\rangle + 0 = 0 + I\langle a, b\rangle$ by the commutative law of addition for integers, $I\langle a, b\rangle + 0 = I\langle a, b\rangle$. We leave the proof of the uniqueness of the additive identity as an exercise for the student.

Definition 7. $-I\langle a, b\rangle = I\langle b, a\rangle$.

Definition 8. *If $I\langle a, b\rangle + I\langle c, d\rangle = 0$ we call $I\langle c, d\rangle$ an* ***additive inverse*** *of $I\langle a, b\rangle$.*

Note that this definition is consistent with our previous definition of $-k$ as equal to $\{(a, b) | (a, b) \sim (1, k + 1)\}$ since we now have $-I\langle k + 1, 1\rangle = I\langle 1, k + 1\rangle$ where $I\langle k + 1, 1\rangle = +k$.

We leave it to the student to prove the following theorem.

Theorem 8. The unique additive inverse of $I\langle a, b\rangle$ is $I\langle b, a\rangle (= -I\langle a, b\rangle)$ and $I\langle b, a\rangle + I\langle a, b\rangle = 0$. Also $-(-I\langle a, b\rangle) = I\langle a, b\rangle$.

We now define *subtraction* of integers.

Definition 9. $I\langle a, b\rangle - I\langle c, d\rangle = I\langle a, b\rangle + (-I\langle c, d\rangle)$.

It is important to note here that the same symbol, " $-$ ", is being used in two quite different ways. This is as customary in elementary algebra, but it can be the source of considerable confusion.† As used in Definition 7 it indicates a *unary* operation that transforms the single integer $I\langle a, b\rangle$ into another integer, $I\langle b, a\rangle$. As used in $I\langle a, b\rangle - I\langle c, d\rangle$, on the other hand, it indicates a binary operation between two integers, $I\langle a, b\rangle$ and $I\langle c, d\rangle$.

† This is recognized in several more recent elementary texts that use, for example, $^{-}2$ for -2 to introduce "signed numbers."

From these definitions, the usual rules of operation with integers follow quickly. We state and prove two such properties here and list others as exercises for the student.

Theorem 9. $-(I\langle a, b\rangle - I\langle c, d\rangle) = -I\langle a, b\rangle + I\langle c, d\rangle$.

Proof. $-(I\langle a, b\rangle - I\langle c, d\rangle) = -[I\langle a, b\rangle + (-I\langle c, d\rangle)] = -(I\langle a, b\rangle + I\langle d, c\rangle) = -I\langle a + d, b + c\rangle = I\langle b + c, a + d\rangle$ by applying the definition of addition of integers together with Definitions 7 and 9. On the other hand, $-I\langle a, b\rangle + I\langle c, d\rangle = I\langle b, a\rangle + I\langle c, d\rangle = I\langle b + c, a + d\rangle$ also.

Theorem 10. $(-I\langle a, b\rangle) \times (-I\langle c, d\rangle) = I\langle a, b\rangle \times I\langle c, d\rangle$.

Proof. $(-I\langle a, b\rangle) \times (-I\langle c, d\rangle) = I\langle b, a\rangle \times I\langle d, c\rangle = I\langle bd + ac, bc + ad\rangle$. On the other hand, $I\langle a, b\rangle \times I\langle c, d\rangle = I\langle ac + bd, ad + bc\rangle$. Since $bd + ac = ac + bd$ and $bc + ad = ad + bc$, the desired conclusion follows.

From now on we will write the integers in the usual way: $\ldots, -2, -1, 0, 1, 2, \ldots$; indicate an arbitrary integer by a single letter such as a, b, x, y etc.; and operate freely with integers as we do in elementary algebra.

Exercises 4.6

1. Complete the proof of Theorem 7.

2. Prove Theorem 8.

3. If x, y, and z represent arbitrary integers prove the following equalities:
(a) $-(x + y) = (-x) - y$
(b) $x(y - z) = xy - xz$
(c) $(-x)y = -(xy) = x(-y)$
(d) $(-x)(y - z) = -(xy) + xz$

PROBLEMS FOR FURTHER INVESTIGATION

1. In the set of all people, the relation "is an uncle of" is clearly connected with the relations "is a brother of" and "is a parent of." Can you state any similar general rule for constructing a new relation out of two given ones? [T1]

2. If R and T are any two relations on a set S can you give reasonable interpretations to $R \subseteq S$, $R = S$, $R \cup S$, and $R \cap S$, ? $\overline{R}$ (See problem 6, Exercises 1.3.) How far can you go in constructing an algebra of relations? [T1]

REFERENCES FOR FURTHER READING

[B4], [E2], [G2], [R2], [T1].

5

Integral Domains

One of the outstanding characteristics of modern mathematics is its tendency to seek common properties of mathematical systems. For example, in this chapter we will define a certain kind of mathematical structure called an integral domain, and we will see that the integers, the rational numbers, the real numbers, and the complex numbers (and other systems) are all examples of integral domains. Thus any theorem that we can prove concerning integral domains will automatically become a theorem concerning the integers, the rational numbers, the real numbers, the complex numbers, etc.

In using as examples of integral domains the rational numbers, the real numbers, and the complex numbers we are not, of course, being systematic since we have only completed the formal development of our number system through the integers. Since, however, the properties of these number systems are familiar to you from your previous work in mathematics, there will be no actual difficulty. And, in this way, we can alternate discussion of the familiar with more general discussion of the less familiar.

1. Definition of an Integral Domain

Definition 1. *Let D be a set containing at least two elements on which are defined two binary operations, $+$ and $\times$, such that* (1) (a) $(a + b) + c = a + (b + c)$ *for all* $a, b, c \in D$ (*associative property of addition*);
(b) $(a \times b) \times c = a \times (b \times c)$ *for all* $a, b, c \in D$.; (*associative property of multiplication*).
(2) $a \times (b + c) = (a \times b) + (a \times c)$ *and* $(b + c) \times a = (b \times a) + (c \times a)$ *for all* $a, b, c \in D$ (*distributive properties*);
(3) *There exists a unique element* $0 \in D$ *such that* $a + 0 = a$ *for all* $a \in D$, (*existence of an additive identity*);
(4) *For every* $a \in D$, *there exists a unique element* $b \in D$ *such that* $a + b = 0$ (*existence of additive inverses*);

(5) (a) $a + b = b + a$ *for all* $a, b \in D$ (*commutative property of addition*);
(b) $a \times b = b \times a$ *for all* $a, b \in D$ (*commutative property of multiplication*);

(6) *There exists a unique element* $1 \in D$ *such that* $1 \times a = a \times 1 = a$ *for all* $a \in D$ (*existence of multiplicative identity*);

(7) *If* $a, b \in D$ *and* $a \times b = 0$, *then* $a = 0$ *or* $b = 0$ (*absence of divisors of zero*).

Such a set is called an ***integral domain.***

As is the case in elementary algebra we write $a \times b$ as ab and denote the additive inverse of a by $-a$. By the commutative property of addition we note that $0 + a = a$ and $(-a) + a = 0$.

Example 1. With the usual definitions of addition and multiplication, it is easy to see that the integers, the rational numbers, the real numbers, and the complex numbers form integral domains. Since we will be using these sets of numbers quite frequently we will find it convenient to establish the following notational conventions:

$I = \{x | x \text{ is an integer}\}$
$R = \{x | x \text{ is a rational number}\}$
$R^* = \{x | x \text{ is a real number}\}$
$C = \{x | x \text{ is a complex number}\}$

Example 2. The set N of natural numbers under the usual definitions of addition and multiplication is not an integral domain (why?) nor is the set $\{0\}$ (why?).

Example 3. Let S be the set of all polynomials† in x with integral coefficients. Then S is an integral domain under the usual definitions of addition and multiplication of polynomials. Here the identity for addition is the polynomial $0(= 0 \cdot x^0)$, the identity for multiplication is the polynomial $1(= 1 \cdot x^0)$ and the additive inverse of $a_0x^n + a_1x^{n-1} + \ldots + a_n$ is $(-a_0)x^n + (-a_1)x^{n-1} + \ldots + (-a_n)$.

To see the significance of Postulate 7 let us consider next an important type of algebraic system—the *residue classes modulo m.*

Definition 2. *Let* a, b, *and* m *be integers with* $m > 0$. *Then* $a \equiv b \pmod{m}$ (*read "a is* ***congruent*** *to b* ***modulo*** *m") if and only if* $a - b$ *is divisible by* m.

We will use the notation $x | y$ for "x divides y" and $x \nmid y$ for "x does not divide y".

† Here, too, we are considering polynomials only in the intuitive sense of elementary algebra, leaving a more precise definition for later.

Examples. $5 \equiv 1 \pmod 4$ since $4|(5-1)$; $16 \equiv 6 \pmod 5$ since $5|(16-6)$; $-5 \equiv -9 \pmod 2$ since $2|[-5-(-9)]$.

Theorem 1. For any fixed integer $m > 0$, the relation of congruence modulo m is an equivalence relation on the integers.

Proof. (1) $a \equiv a \pmod m$ since $a - a = 0$ and $m|0$.

(2) If $a \equiv b \pmod m$, then $m|(a-b)$. But then $m|(b-a)$ and hence $b \equiv a \pmod m$.

(3) If $a \equiv b \pmod m$ and $b \equiv c \pmod m$, then $m|(a-b)$ and $m|(b-c)$. Thus $m|[(a-b)+(b-c)]$ so that $m|(a-c)$ and hence $a \equiv c \pmod m$.

Since $\equiv$ is an equivalence relation we know, by Theorem 4.1, that it partitions the integers. For example, if $m = 5$, the partitioning classes are:

$$\begin{array}{l}\{\ldots, -10, -5, 0, 5, 10, \ldots\} \\ \{\ldots, -9, -4, 1, 6, 11, \ldots\} \\ \{\ldots, -8, -3, 2, 7, 12, \ldots\} \\ \{\ldots, -7, -2, 3, 8, 13, \ldots\} \\ \{\ldots, -6, -1, 4, 9, 14, \ldots\}\end{array}$$

In general, for given m, we denote these classes by $C_0, \ldots, C_{m-1}$ where

$$C_i = \{a | a \in I \text{ and } a \equiv i \pmod m\}, \ i = 0, 1, \ldots, m-1.$$

Each one of these C_i is called a *residue class* and the set $\{C_0, \ldots, C_{m-1}\}$ will be denoted by J_m.

We now define addition and multiplication in J_m.

Definition 3. (1) $C_i + C_j = \{a | a \in I \textit{ and } a \equiv i + j \pmod m\}$
(2) $C_i C_j = \{a | a \in I \textit{ and } a \equiv ij \pmod m\}$

Examples. In J_5 we have $C_1 + C_2 = C_3$, $C_4 + C_3 = C_2$, $C_1C_2 = C_2$, $C_3C_4 = C_2$ since $1 + 2 \equiv 3 \pmod 5$, $4 + 3 \equiv 2 \pmod 5$, $1 \cdot 2 \equiv 2 \pmod 5$, and $3 \cdot 4 \equiv 2 \pmod 5$.

For brevity we frequently write 0 for C_0, 1 for $C_1, \ldots, k$ for C_k. Thus, for example, we consider $J_3 = \{0, 1, 2\}$ with $1 + 1 = 2$, $1 + 2 = 0$, $2 \times 1 = 2$, $2 \times 2 = 1$ etc.

Exercises 5.1

In problems 1–10 determine whether or not the given sets are integral domains with respect to the usual definitions of addition and multiplication.

1. $\{b\sqrt{2} | b \in R\}$
2. $\{3m | m \in I\}$
3. $\{a + b\sqrt{2} | a, b \in I\}$
4. $\{a + b\sqrt{2} | a, b \in R\}$
5. $\{a + bi \ | a, b \in R\}$
6. $\{a + b\sqrt[3]{9} | a, b \in R\}$

7. The set of all polynomials $f(x)$ with integral coefficients such that $f(-x) = f(x)$

8. $\{a + b\sqrt[3]{2} + c\sqrt[3]{4} | a, b, c \in R\}$

9. The set of even integers

10. $\{1\}$

11. Construct addition and multiplication tables for J_3, J_4, and J_5

***12.** Show that the word "unique" may be omitted in Postulates 3, 4, and 6 in Definition 1; that is, prove the uniqueness from the assumption of the existence of at least one such element as described.

2. Elementary Properties of Integral Domains

Many of the most important properties of integral domains depend only upon closure, the commutative law of addition, the associative laws, the distributive laws, the existence of an additive identity, and the existence of additive inverses. For this reason it is useful to make the following definition.

Definition 4. *A set R on which are defined two binary operations, $+$ and $\times$, satisfying properties* 1 *to* 5a *of Definition* 1 *is called a* ***ring****. A ring which also satisfies property* 5b *is called a* ***commutative*** *ring.*

We will discuss rings in more detail in Chapter 9. Here we prove only a few simple properties.

All integral domains are, of course, commutative rings and hence any result proved for rings or commutative rings will hold for integral domains. Some examples of commutative rings that are not integral domains are:

1. The set of even integers (no multiplicative identity)
2. The set consisting of 0 alone (does not have at least two elements)
3. J_4 (does not satisfy postulate 7 since $2 \times 2 = 0$ in J_4 and $2 \neq 0$).

Examples of noncommutative rings will be given later.

Definition 5. *If $a \neq 0$ and $b \neq 0$ in a ring R but $a \times b = 0$ we call a and b* ***divisors of zero****.*

Thus 2 and 3 are both divisors of zero in J_6 since $2 \times 3 = 0$ but 5 is not a divisor of zero since $5 \times b = 0$ if and only if $b = 0$.

Our main results concerning rings are given in the following theorem.

Theorem 2. In any ring R we have
(1) $b + a = c + a$ implies $b = c$ (cancellation law for addition);
(2) The equation $x + a = c$ has the unique solution $x = c + (-a)$;
(3) $-(-a) = a$;
(4) $a \times 0 = 0 \times a = 0$;
(5) $(-a)(-b) = ab$;
(6) $-(ab) = (-a)b = a(-b)$.

Proof of (1). From $b + a = c + a$ we have $(b + a) + (-a) = (c + a) + (-a)$. Re-associating we have $b + [a + (-a)] = c + [a + (-a)]$. Since $a + (-a) = 0$ we have $b + 0 = b = c + 0 = c$.

Proof of (2). That $x = c + (-a)$ is a solution follows from $[c + (-a)] + a = c + [(-a) + a] = c + 0 = c$. The uniqueness follows from the cancellation law since if $x + a = c$ and $x' + a = c$, then $x + a = x' + a$ and $x = x'$.

Proof of (3). Since $(-a) + a = 0$ it follows that a is the additive inverse of $-a$; i.e., $-(-a) = a$.

Proof of (4). $(a \times 0) + (a \times a) = a \times (0 + a) = a \times a = 0 + (a \times a)$. From $(a \times 0) + (a \times a) = 0 + (a \times a)$ we obtain $a \times 0 = 0$ by the cancellation law. The proof that $0 \times a = 0$ is similar and is left as an exercise for the student.

Proof of (5). We have $(-a)(-b) + [(-a)b + ab] = (-a)(-b) + [(-a) + a]b = (-a)(-b) + 0 \times b = (-a)(-b) + 0 = (-a)(-b)$. But, also, $(-a)(-b) + [(-a)b + ab] = [(-a)(-b) + (-a)b] + ab = (-a)[(-b) + b] + ab = (-a) \times 0 + ab = 0 + ab = ab$. Therefore, $(-a)(-b) = ab$. (The student should supply reasons for each step.)

Proof of (6). By definition of $-(ab)$ it is the unique solution to the equation $ab + x = 0$. But, also, $ab + (-a)b = [a + (-a)]b = 0 \times b = 0$ so that $(-a)b$ is also a solution to $ab + x = 0$. Hence $(-a)b = -(ab)$. The proof that $-(ab) = a(-b)$ is left as an exercise for the student.

Theorem 3. Suppose D is an integral domain and $a, b, c \in D$. Then $ab = ac$ and $a \neq 0$ implies that $b = c$ (cancellation law for multiplication).

Proof. If $ab = ac$, $ab + [-(ac)] = ac + [-(ac)] = 0$. But $-(ac) = a(-c)$ by Theorem 2 and hence $ab + [-(ac)] = ab + a(-c) = a[b + (-c)] = 0$. Therefore, since $a \neq 0$, the absence of divisors of zero demands that $b + (-c) = (-c) + b = 0$. Hence $b = -(-c) = c$.

Definition 6. *In any ring R we define, for $a, b \in R$, $a - b = a + (-b)$.*

Exercises 5.2

1. Determine which of the sets in problems 1–10 of Exercises 5.1 are rings.
2. Prove that J_m is a ring for every $m \in N$.
3. Consider the set S of pairs of rational numbers (a, b) with equality, addition, and multiplication defined as follows: $(a, b) = (c, d)$ if and only if $a = c$ and $b = d$; $(a, b) + (c, d) = (a + c, b + d)$; $(a, b) \times (c, d) = (ac, bd)$. Prove that S is a ring but not an integral domain.

4. In any ring S prove that, for $x, y, z \in S$,
(a) $-(x + y) = (-x) - y$
(b) $x(y - z) = xy - xz$
(c) $(-x)(y - z) = -(xy) + xz$
(d) $-(x - y) = -x + y$

5. If a ring R has a unity element, 1, prove that $(-1)a = -a$ for all $a \in R$.

6. Complete the proof of (4) of Theorem 2.

7. Complete the proof of (6) of Theorem 2.

***8.** Let R be the set of all subsets of a given set S. For $A, B \in R$ we define $A + B = \{x | x \in A \text{ or } x \in B \text{ but } x \notin A \cap B\}$ and $A \times B = A \cap B$. Prove that R is a ring under these two operations.

3. Division in an Integral Domain

Definition 7. *Let D be an integral domain with $a, b, c \in D$. Then if $ab = c$ we say that a* **divides** *c or that a is a* **divisor** *of c and write $a|c$.*

Definition 8. *If D is an integral domain, $a \in D$, and $a|1$ we say that a is a* **unit** *of D.*

Clearly in any integral domain the multiplicative identity, 1, and its additive inverse, -1, are units.

Example 1. If $D = I$ it is obvious that the only units of D are 1 and -1.

Example 2. Let $D = \{a + bi | a, b \in I\}$ $(i = \sqrt{-1})$. It is easy to show that D is an integral domain and that the units of D are $1, -1, i$, and $-i$.

Example 3. Let $D = R$. Then since for every nonzero rational number a/b, $1 \div (a/b) = b/a$, every nonzero rational number is a unit of R.

Definition 9. *Let D be an integral domain with $a, b \in D$. We call a and b* **associates** *if $a = bu$ where u is a unit of D and $ab \neq 0$.*

Example 4. In the integral domain of Example 2, $5, -5, 5i, -5i$ are the four associates of 5.

Example 5. If $D = R$ every number except 0 is a unit (Example 3) and hence any two nonzero numbers are associates.

Theorem 4. If D is an integral domain and u and u' are units of D, then uu' is a unit of D.

Proof. $u|1$ and $u'|1$. Thus there exist $c, c' \in D$ such that $uc = 1$ and $u'c' = 1$. Hence $(uc)(u'c') = (uu')(cc') = 1$ and $uu'|1$.

Theorem 5. Define $a\,R\,b$ if and only if a and b are associates. Then R is an equivalence relation on the set of nonzero elements of an integral domain D.

Proof. (1) $a\,R\,a$ since $a = a \cdot 1$ and 1 is a unit of any integral domain. Hence R is reflexive.

(2) If $a\,R\,b$, then $a = bu$ where u is a unit of D. Hence there exists $c \in D$ such that $uc = cu = 1$. Thus $c|1$ and c is also a unit of D. But $ac = (bu)c = b(uc) = b \cdot 1 = b$ so that b is equal to a times a unit of D. Thus $b\,R\,a$ and hence R is symmetric.

(3) If $a\,R\,b$ and $b\,R\,c$ we have $a = bu$, $b = cu'$ where u and u' are units of D. Then $a = bu = (cu')u = c(u'u)$. Since $u'u$ is a unit of D by Theorem 4 it follows that $a\,R\,c$ and hence that R is transitive.

Definition 10. *If D is an integral domain and $a \in D$ we say that a is a* ***prime*** *or* ***irreducible*** *element of D if a is not a unit and if $a = bc$ implies b or c is a unit. If a is neither a prime nor a unit we say that it is* ***composite*** *or* ***reducible****.*

Example 6. In I the prime elements are the ordinary primes of arithmetic, $(\pm 2, \pm 3, \pm 5, \ldots)$.

Example 7. It is easy to verify that the set of all polynomials with integral coefficients is an integral domain D and that the units of D are 1 and -1. Then, for example, $x^2 + x + 1$, $2x + 5$, and 7 are all prime elements of D but $x^2 + 5x + 6\ [= (x + 2)(x + 3)]$, $2x + 6[= 2(x + 3)]$ and $8(= 2 \cdot 4)$ are not.

Example 8. Let $D = \{a + b\sqrt{5} | a, b \in I\}$. We leave it to the student to verify that D is an integral domain and now seek to determine the units of D. To do this it is convenient to introduce what is called a *norm* in D.† To this end suppose that $\alpha = a + b\sqrt{5} \in D$. Then $N(\alpha)$, the norm of α, is defined by

$$N(\alpha) = a^2 - 5b^2 \in I$$

If, now, $\beta = c + d\sqrt{5} \in D$, direct computation shows that $N(\alpha\beta) = N(\alpha)N(\beta)$. Now if $\alpha = a + b\sqrt{5}$ is a unit, then there exists $\beta = c + d\sqrt{5} \in D$ such that $\alpha\beta = 1$ and hence $N(\alpha\beta) = N(1) = 1$. Thus we must have

$$N(\alpha\beta) = N(\alpha)N(\beta) = (a^2 - 5b^2)(c^2 - 5d^2) = 1$$

† This will indeed be useful. Unfortunately, not all integral domains have useful norms.

Hence if α is a unit we must have $(a^2 - 5b^2)(c^2 - 5d^2) = 1$. Thus $N(\alpha) = a^2 - 5b^2 = \pm 1$ if $\alpha = a + b\sqrt{5}$ is a unit. On the other hand, if $N(\alpha) = 1$, then $(a + b\sqrt{5})(a - b\sqrt{5}) = a^2 - 5b^2 = 1$ and if $N(\alpha) = -1$, then $(a + b\sqrt{5})(-a + b\sqrt{5}) = -(a^2 - 5b^2) = 1$. Thus if $N(\alpha) = \pm 1$, $\alpha | 1$ so that α is a unit of D.

We have thus shown that $\alpha = a + b\sqrt{5}$ is a unit of D if and only if $N(\alpha) = \pm 1$. From this fact we conclude that

(1) $1, -1, 2 - \sqrt{5}, 2 + \sqrt{5}, -2 - \sqrt{5}, -2 + \sqrt{5}$

are all units of D and that

(2) $1 + \sqrt{5}$ and $-3 + \sqrt{5}$

are not units of D. Furthermore, since $(2 - \sqrt{5})(1 + \sqrt{5}) = -3 + \sqrt{5}$,

(3) $1 + \sqrt{5}$ and $-3 + \sqrt{5}$

are associates.

We may now prove that

(4) $2, 1 + \sqrt{5}$, and $-1 + \sqrt{5}$

are all primes of D. For if $2 = \alpha\beta$ where neither α nor β are units, $N(2) = 4 = N(\alpha)N(\beta)$ means that $N(\alpha) = \pm 2$ and $N(\beta) = \pm 2$. But if $N(\alpha) = a^2 - 5b^2 = \pm 2$, then $a^2 \equiv \pm 2 \pmod 5$ and it is easy to see that there is no solution to this congruence. (Simply try $a = 0, 1, -1, 2$ and -2 in turn.) We leave it to the student to prove that $1 + \sqrt{5}$ and $-1 + \sqrt{5}$ are also primes of D. Furthermore 2 is not an associate of $1 + \sqrt{5}$ since if $2 = (a + b\sqrt{5})(1 + \sqrt{5})$ where $a + b\sqrt{5}$ is a unit of D, we have $2 = (a + 5b) + (a + b)\sqrt{5}$. Hence $a + 5b = 2$ and $a + b = 0$. Thus $a = -b$ and $4b = 2$. It follows that $b = 1/2$, a contradiction to $a + b\sqrt{5} \in D$.

Now we observe that

(5) $4 = 2 \cdot 2 = (1 + \sqrt{5})(-1 + \sqrt{5})$

so that we have two factorizations of 4 in D into primes that are not associates. This is in contrast to factorization into primes in I where the factorization is unique (except for units)† Thus we see that we do not have unique factorization into primes in every integral domain.

Exercises 5.3

1. What are the units of J_5?

2. Prove that the only units of $\{a + bi | a, b \in I\}$ are $1, -1, i$, and $-i$.

† A formal proof of this will be given in Section 7.

3. Prove that u is a unit of an integral domain D if and only if u has a multiplicative inverse in D.
4. Prove that a and b are associates in an integral domain D if and only if $a|b$ and $b|a$.
5. In Example 8, prove that $N(\alpha\beta) = N(\alpha)N(\beta)$.
6. In Example 8, prove that $1 + \sqrt{5}$ and $-1 + \sqrt{5}$ are primes.
7. Modify, if necessary, the definitions of this section so that they apply to commutative rings with an identity.
8. Do as in problem 7 for arbitrary (possibly noncommutative) rings with an identity.
9. Let $D = \{a + b\sqrt{13}|a, b \in I\}$. Prove that there is not unique factorization into primes in D.
10. Show that there are an infinite number of units in $\{a + b\sqrt{2}|a, b \in I\}$.

4. Ordered Integral Domains

Definition 11. *An integral domain D is said to be* **ordered** *if there exists a subset P of the elements of D such that*
(1) if $a, b \in P$, then $a + b \in P$ (closure under addition);
(2) if $a, b \in P$, then $ab \in P$ (closure under multiplication);
(3) if $a \in D$ then one and only one of the alternatives $a \in P$, $a = 0$, and $-a \in P$ hold (the trichotomy law).
The set P is called a set of **positive elements** *of D.*

Example 1. Let $D = I$ and P the set of all (ordinary) positive integers ($= N$). (Keep in mind that here, as in similar situations before, we must distinguish between the formal concept and an informal example. In Definition 11 all we know about the concept of positive element is what is contained in the definition. In our examples, however, we are using our intuitive notion of positive ("ordinary") to see that it agrees with our formal notion.)

Example 2. Let $D = R$ and P be the set of all positive rational numbers.

Example 3. The integral domain, J_3, of integers modulo 3 is not an ordered integral domain.

Proof. We may take the elements of J_3 to be 0, 1, and 2. Suppose P is a set of positive elements of J_3. We note first that $0 \notin P$ by the trichotomy law. Now if $1 \in P$, $1 + 1 = 2 \in P$ by (1) of Definition 11 and hence $1 + 2 = 0 \in P$ contrary to fact. Thus $1 \notin P$; hence by (3), $-1 = 2 \in P$ since $1 \neq 0$. But then $2 + 2 = 1 \in P$, again a contradiction.

Example 4. The integral domain, C, of complex numbers is not an ordered integral domain.

Proof. If $i \in P$, then $i^2 = -1 \in P$ by (2) of Definition 11. But then $(-1)(-1) = 1 \in P$ so that we have $1 \in P$ and also $-1 \in P$ contrary to (3). Thus $i \notin P$ and, since $i \neq 0$, $-i \in P$ by (3). But then we have $(-i)(-i) = -1 \in P$ and again obtain a contradiction.

We may ask whether or not it is possible to define more than one set of positive elements for a given integral domain. Example 5 will show that this is indeed possible. If, however, $D = I$, only one definition is possible. More precisely, we have

Theorem 6. If $D = I$, then $P = N$.

Proof. $1 \in P$ since if $1 \notin P$, $-1 \in P$ by (3). But then $(-1)(-1) = 1 \in P$ by (2), a contradiction. Then if n is any natural number,

$$n = \overbrace{1 + 1 + 1 + \ldots + 1}^{n \text{ terms}} \in P \text{ by (1)}$$

But then $-n \notin P$ by (3) and, of course, $0 \notin P$.

Example 5. Let $D = \{a + b\sqrt{2} \mid a, b \in I\}$. It is easy to show that D is an integral domain. Clearly $D \subset R^*$ and we suppose that we have an intuitive understanding of the "natural" (and actually unique) ordering for the real numbers in the sense that numbers are called positive if and only if their representation on the real number line is to the right of the origin. Let P be the set of positive elements of R^* in this sense. We now define a set of positive elements, P_1, in D by

(1) $a + b\sqrt{2} \in P_1$ if and only if $a + b\sqrt{2} \in P$

and another set of positive elements, P_2, in D by

(2) $a + b\sqrt{2} \in P_2$ if and only if $a - b\sqrt{2} \in P$

(e.g., $5 - 6\sqrt{2} \in P_2$ since $5 - (-6)\sqrt{2} = 5 + 6\sqrt{2} \in P$).

Now (1) certainly defines an ordering in D since D is a subset of R^* whose natural ordering we are accepting. To show that (2) also defines an (obviously different) ordering we observe that if $a + b\sqrt{2}$, $c + d\sqrt{2} \in P_2$, then $a - b\sqrt{2}$, $c - d\sqrt{2} \in P$. Hence

$$(a - b\sqrt{2}) + (c - d\sqrt{2}) = (a + c) - (b + d)\sqrt{2} \in P$$

and therefore

$$(a + c) + (b + d)\sqrt{2} = (a + b\sqrt{2}) + (c + d\sqrt{2}) \in P_2$$

Similarly

$$(a - b\sqrt{2})(c - d\sqrt{2}) = (ac + 2bd) - (ad + bc)\sqrt{2} \in P$$

and therefore

$$(ac + 2bd) + (ad + bc)\sqrt{2} = (a + b\sqrt{2})(c + d\sqrt{2}) \in P_2$$

Thus we see that conditions (1) and (2) of Definition 11 are satisfied. Finally, we observe that either

$$a - b\sqrt{2} = 0 \text{ (whence } a = b = 0), a - b\sqrt{2} \in P,$$
$$\text{or} \quad -(a - b\sqrt{2}) = -a + b\sqrt{2} \in P$$

and that only one of these alternatives may hold by our assumption that P is a set of positive elements in $R^* \supset D$. Hence
$a + b\sqrt{2} = 0$, $a + b\sqrt{2} \in P_2$, or $-a - b\sqrt{2} = -(a + b\sqrt{2}) \in P_2$ by our definition of $a + b\sqrt{2} \in P_2$. Thus condition (3) of Definition 11 is satisfied.

Definition 12. *Let D be an ordered integral domain and P a set of positive elements of D. Then by*
(1) $a > 0$ we mean $a \in P$
(2) $a > b$ we mean $a - b > 0$
(3) $a < b$ we mean $b > a$.
where $a, b \in D$.

(NOTE: Technically, we should write $a >_P 0$ etc., since the meaning of "$>$" depends on P. Thus, in Example 5, $1 + \sqrt{2} >_{P_1} 0$ but $1 + \sqrt{2} \ngtr_{P_2} 0$ since $1 - \sqrt{2} \notin P$. We shall understand, however, that we are always referring to a fixed P whenever we use the symbol "$>$" in connection with a given integral domain.)

Theorem 7. Let D be an ordered integral domain. Then, for all $a, b, c \in D$
(1) If $a > b$ and $b > c$, then $a > c$ (transitive property)
(2) If $a > b$, then $a + c > b + c$
(3) For any a and b, one and only one of the relations $a > b$, $a = b$, or $a < b$ hold
(4) If $a > b$ and $c > 0$, then $ac > bc$.

Proof of (1). If $a > b$ and $b > c$ we have $a - b \in P$ and $b - c \in P$ by Definition 12. Hence $(a - b) + (b - c) = a - c \in P$ by Definition 11. Thus $a > c$ by Definition 12.

The proofs of the other properties are left as exercises.

Definition 13. *$a \geq b$ if and only if $a > b$ or $a = b$; $a \leq b$ if and only if $a < b$ or $a = b$.*

Theorem 8. Let D be an ordered integral domain. Then, for all a, b, $c \in D$
(1) If $a \geq b$ and $b \geq c$, then $a \geq c$
(2) If $a \geq b$ then $a + c \geq b + c$
(3) If $a \geq b$ and $c \geq 0$, then $ac \geq bc$
(4) If $a \geq b$ and $c \geq d$, then $a + c \geq b + d$.

The proof of Theorem 8 is left as an exercise for the student.

Definition 14. *Suppose $D' \supset D$ where D' and D are both ordered integral domains with sets of positive elements P' and P respectively. We will say that the ordering defined by P' is an* **extension** *of the ordering defined by P if and only if $P = P' \cap D$ (that is, the positive elements of D remain positive when considered as elements of D' and conversely, the positive elements of D' which are elements of D are also positive elements of D).*

Example 6. In Example 5 the natural ordering of the real numbers is an extension of the ordering in D defined by P_1 but is not an extension of the ordering in D defined by P_2.

In an ordered integral domain, the concept of absolute value is a useful one.

Definition 15. *If D is an ordered integral domain with P a set of positive elements of D we define the* **absolute value,** *$|a|$, of $a \in D$ by $|a| = a$ if $a = 0$ or $a \in P$ and $|a| = -a$ if $-a \in P$.*

Thus if D is the set of rational numbers, $|2| = 2$, $|0| = 0$, $|-1/2| = -(-1/2) = 1/2$. Again (see note following Definition 12), the concept of absolute value in an integral domain depends on P and we should really write $|a|_P$. As in the use of "$>$," however, we will assume that we are always referring to a fixed P when we use the absolute value symbol.

Theorem 9. If D is an ordered integral domain and $a, b \in D$, then (1) $|ab| = |a||b|$ and (2) $|a + b| \leq |a| + |b|$.

Proof. The proof of (1) is left as an exercise for the student. To prove (2) observe that if either (a) $a > 0$ and $b > 0$, (b) $a < 0$ and $b < 0$, or (c) $a = 0$ or $b = 0$, we have $|a + b| = |a| + |b|$. Now if $a > 0$ and $b < 0$ we have $|a| = a$ and $|b| = -b$. Then if $a + b < 0$, $|a + b| = -(a + b)$. But, since $a > 0$, $-(a+b) = -a - b < a - b = a + (-b) = |a| + |b|$. If $a + b \geq 0$, $|a + b| = a + b$ and $a + b < a - b = |a| + |b|$. We merely interchange the roles of a and b in the above argument if $a < 0$ and $b > 0$.

Exercises 5.4

1. Prove that in an ordered integral domain, D, the multiplicative identity of D is a positive element of D.
2. Complete the proof of Theorem 7.
3. Prove Theorem 8.
4. Let D be an ordered integral domain with $a, b, c \in D$. Prove that
   (a) If $a \geq b$ and $c < 0$, then $ac \leq bc$
   (b) If $a \geq b \geq 0$ and $c \geq d \geq 0$, then $ac \geq bd$.
5. Prove (1) of Theorem 9.

5. Variations of the Principle of Mathematical Induction

Now that we have available the integers and concepts of inequality and divisibility for the integers we can formulate useful variants of the principle of mathematical induction. Our most important variant is the well-ordering principle which we will establish by first proving three lemmas.

Lemma 1. $n \geq 1$ for all $n \in N$.

Proof. Let $S = \{n | n \in N \text{ and } n \geq 1\}$. Clearly $1 \in S$ since $1 \geq 1$. Now suppose $k \in S$ so that $k \geq 1$. From $k \geq k$ and $1 \geq 0$ we obtain $k + 1 \geq k + 0 = k$ so that $k + 1 \geq 1$ follows by transitivity. Hence $k \in S$ implies $k + 1 \in S$ so that, by the principle of mathematical induction, $S = N$.

Lemma 2. If $a \in N$, then $\{n | n \in N \text{ and } a + 1 > n > a\} = \varnothing$.

Proof. Suppose that there is a natural number n such that $a + 1 > n > a$. Then $(a + 1) - n > 0$ and $n - a > 0$ so that $(a + 1) - n = x \in N$ and $n - a = y \in N$ by Definition 12 and Theorem 6. Hence $[(a + 1) - n] + (n - a) = 1 = x + y$. But, by Lemma 1, $x \geq 1$ and $y \geq 1$ so that, by (4) of Theorem 8, $x + y \geq 1 + 1 \neq 1$. We thus have a contradiction to our assumption that there exists a natural number n such that $a + 1 > n > a$ and hence $\{n | n \in N \text{ and } a + 1 > n > a\} = \varnothing$.

Lemma 3. If h and k are two natural numbers such that $k + 1 > h$, then $k \geq h$.

Proof. Since $k + 1 > h$, it follows that $(k + 1) - h > 0$ so that $k + 1 - h = x \in N$ by Definition 12 and Theorem 6. Now $x \geq 1$ by Lemma 1 and if $x = 1$, $k + 1 - h = 1$. Hence $k - h = 0$ and $k = h$. On the other hand, if $x > 1$, then $x - 1 > 0$ and $x - 1 = y \in N$. Hence $x = y + 1$ so that $k + 1 - h = y + 1$ and $k - h = y \in N$. Thus $k - h > 0$ and $k > h$.

Theorem 10. (*the well-ordering principle*) In every nonempty set S of natural numbers there exists a smallest one; i.e. there exists $x \in S$ such that $y \in S$ implies $y \geq x$.

Proof. Our proof will be made by assuming that S does not have a smallest, element and obtaining a contradiction. Under this assumption, then Lemma 1 tells us that $1 \notin S$. We define $T = \{n | n \in N \text{ and } n < s \text{ for all } s \in S\}$. Clearly $S \cap T = \varnothing$ and $1 \in T$.

We proceed as follows:

(1) We show that $k \in T$ implies $k + 1 \notin S$ by showing that if $k + 1 \in S$ it would be the smallest element in S, contrary to our assumption that S has no smallest element;

(2) We show that if $k \in T$, then $k + 1 \in T$.

Then, since $1 \in T$, (2) and the principle of mathematical induction gives us that $T = N$. Hence we would have $T \cap S = S \neq \varnothing$ so that our assumption that S has no smallest element is contradicted.

To prove (1) we first observe that if $y \in T$ and $x < y$ for $x \in N$, then $x \in T$. For we have, for any $s \in S$, $y < s$ by definition of T, and since $x < y$ it follows that $x < s$ for all $s \in S$ by (1) of Theorem 7 (transitive law); i.e. $x \in T$. Now we suppose that $k \in T$ but $k + 1 \in S$. Then if there exists $h \in S$ such that $h < k + 1$ (that is, if $k + 1$ is not the smallest member of S) we have $h \leq k$ by Lemma 3. But $h \neq k$ since $h \in S$, $k \in T$, and $S \cap T = \varnothing$. Thus $h < k \in T$ so that, by our previous remarks, $h \in T$. However, $h \in S$ and $S \cap T \neq \varnothing$. a contradiction. Hence there is no element $h \in S$ such that $h < k + 1$ and $k + 1$ is the smallest member of S, a contradiction to our assumption that S has no smallest element.

(2) We have shown that for any $k \in T$, we have $k + 1 \notin S$. But this implies that $k + 1 \in T$ for if $k + 1 \notin T$ there exists $s \in S$ with $k + 1 \geq s > k$. But if $k + 1 = s$, then $k + 1 \in S$, contrary to (1). Hence we have $k + 1 > s > k$, a contradiction to Lemma 2. Thus $k + 1 \in T$.

Theorem 11. Let $S \subseteq N$. Then if (1) $1 \in S$ and (2) $m \in S$ if $k \in S$ for all $k < m$, it follows that $S = N$.

Proof. Let $M = \{t | t \in N \text{ and } t \notin S\}$. We wish to show that $M = \varnothing$ and thus that $S = N$. Now by Theorem 10, if $M \neq \varnothing$ it contains a smallest member, say m, and, by (1), $m \neq 1$. Then for $k \in N$ and $k < m$ we have $k \notin M$ so that $k \in S$ by definition of M. But then $m \in S$ by (2), a contradiction to $m \in M$. Hence our assumption that $M \neq \varnothing$ is false; $M = \varnothing$ and $S = N$.

Example 1. Prove that every natural number greater than 1 is divisible by some prime number.

We let $S = \{1\} \cup \{n | n \in N \text{ and } n \text{ divisible by a prime}\}$. Obviously, $1 \in S$. Suppose $k \in S$ for all $k < m$. Then if m is a prime we are done since $m | m$. If m is not a prime it has a divisor $s \neq 1$ and $m = st$. But then, since $s < m$, our induction hypothesis that $k \in S$ for all $k < m$, says that s has a prime divisor, say p. Hence $s = pr$ and $m = (pr)t = p(rt)$ so that $p | m$. By Theorem 11, $S = N$.

Note that here we cannot use the original principle of induction since we know nothing of the relation between the divisors of k and the divisors of $k + 1$. Thus assuming $k \in S$ alone does not help us to prove $k + 1 \in S$.

Theorem 12. Let $S \subseteq N$. Then if (1) $m \in S$ and (2) for all $k \geq m$, $k \in S$ implies $k + 1 \in S$, then $S \supseteq \{n | n \in N \text{ and } n \geq m\}$.

Proof. We let $S' = \{1, 2, \ldots, m\} \cup S$ so that if we show that $S' = N$ it will follow that $S \supseteq \{n | n \in N \text{ and } n \geq m\}$. Obviously $1 \in S'$. Now suppose $k \in S'$. By Lemma 1 we have $k \geq 1$. Further, either $k = m - 1$, $k > m - 1$, or $k < m - 1$. If $k = m - 1$, then $k + 1 = m \in S'$. If $k = (k - 1) + 1 > m - 1$, then, by Lemma 3, $k - 1 \geq m - 1$ so that $k \geq m$. But then $k + 1 \in S$ by (2) and hence $k + 1 \in S'$. Finally, if $k < m - 1$ we have $k \leq (m - 1) - 1 = m - 2$ by Lemma 3 so that $1 \leq k \leq m - 2$. But then $1 + 1 = 2 \leq k + 1 \leq (m - 2) + 1 = m - 1$ so that $k + 1 \in \{1, 2, \ldots, m\}$ and, again, $k + 1 \in S'$. Since $1 \in S'$ and $k \in S'$ implies $k + 1 \in S'$ we conclude that $S' = N$ as desired.

Let us now consider some further examples of proofs by mathematical induction.

Example 2. Prove that $n^3 + 1 > n^2 + n$ for all $n \in N$ with $n \geq 2$.

We let $S = \{n | n \in N \text{ and } n^3 + 1 > n^2 + n\}$ and wish to show that $S \supseteq \{n | n \in N \text{ and } n \geq 2\}$. Now $2 \in S$ since $2^3 + 1 > 2^2 + 2$. Take $k \geq 2$. If $k \in S$, then

(1) $k^3 + 1 > k^2 + k$.

On the other hand, $k + 1 \in S$ if and only if

(2) $(k + 1)^3 + 1 > (k + 1)^2 + (k + 1)$.

To establish (2) we add $3k^2 + 3k + 1$ to both sides of (1) and obtain

$$(k^3 + 3k^2 + 3k + 1) + 1 = (k + 1)^3 + 1 > 4k^2 + 4k + 1 = \\ = (k + 1)^2 + (k + 1) + (3k^2 + k - 1)$$

Now $3k^2 + k = k(3k + 1) > 1$ since $k > 1$ and $3k + 1 > 1$. Thus $3k^2 + k - 1 > 0$ and $(k + 1)^2 + (k + 1) + (3k^2 + k - 1) > (k + 1)^2 + (k + 1)$. Hence $(k + 1)^3 + 1 > (k + 1)^2 + (k + 1)$ which is (2). Thus $k + 1 \in S$ and, by Theorem 12, S contains the set of all natural numbers ≥ 2. Since $1 \notin S$, $S = \{n | n \in N \text{ and } n \geq 2\}$.

Example 3. Prove that the maximum number of lines determined by n points in a plane is $n(n - 1)/2$.

Let $S = \{n | n$ is a natural number and the maximum number of lines determined by n points is $n(n - 1)/2\}$. Now $1 \in S$ since, if $n = 1$, $n(n - 1)/2 = 0$ and one point does not *determine* a line. (Since this is a rather exceptional case we had better see if $2 \in S$. Now $2 \in S$ since 2 points do determine exactly one line and, if $n = 2$, $2(2 - 1)/2 = 1$.)

Now suppose $k \in S$. That is, we suppose that k points determine at most $k(k - 1)/2$ lines. Now suppose that we have a $(k + 1)st$ point. We can draw at *most* k new lines joining the new point with the previous k points. (Some of these lines may coincide with previous lines as shown below for $k = 2$.)

1 2 3

Thus the maximum number of lines determined by $k + 1$ points would be $[k(k - 1)/2] + k = (k + 1)[(k + 1) - 1]/2$ which, for $n = k + 1$, is $n(n - 1)/2$. Thus if $k \in S$, then $k + 1 \in S$ and hence $S = N$.

Example 4. Prove that, for any $n \in N$, $64|3^{2n+2} - 8n - 9$. We let $S = \{n|n \in N \text{ and } 64|3^{2n+2} - 8n - 9\}$. Then $1 \in S$ since if $n = 1$, $3^{2n+2} - 8n - 9 = 3^4 - 8 - 9 = 64$ and $64|64$. Now suppose $k \in S$ so that $64|3^{2k+2} - 8k - 9$. Then $3^{2k+2} - 8k - 9 = 64q$ for some integer q. Consider $3^{2(k+1)+2} - 8(k + 1) - 9 = 3^{2k+4} - 8k - 17 = 3^2(3^{2k+2} - 8k - 9) + 64k + 64 = 9(64q) + 64k + 64 = 64(9q + k + 1)$. Hence $64|3^{2(k+1)+2} - 8(k + 1) - 9$ and $k + 1 \in S$. By the principle of mathematical induction, $S = N$.

Exercises 5.5

In problems 1–11 use mathematical induction to establish the stated results.

1. If $a > -1$, then $(1 + a)^n \geq 1 + na$ for all $n \in N$.
2. The maximum number of intersections of n lines in a plane is $n(n - 1)/2$ for all $n \in N$, $n \geq 2$.
3. The number of regions into which n lines divide the plane can never exceed 2^n.
4. Let D be an ordered integral domain with a set of positive elements P. Then if $-a \in P$ it follows that $-a^{2n+1} \in P$ for all $n \in N$.
5. $5|8^n - 3^n$ for all $n \in N$.
6. $64|9^n - 8n - 1$ for all $n \in N$.
7. $2304|7^{2n} - 48n - 1$ for all $n \in N$.
8. $5|2^{4n} - 1$ for all $n \in N$.
9. If $n \in N$, then $n < 2^n$.
10. If $n \in N$, then $2^{n+3} < (n + 3)!$.
11. If $n, r \in N$, then $n!r! < (n + r)!$.
12. Show that for any integer a, $a - 1$ is the largest integer less than a.
13. A subset S of an ordered integral domain is said to be *well ordered* if every nonempty subset of S contains a smallest member.
    (a) Which of the following sets are well ordered? $\{2n + 1|n \in N\}$; $\{-2n|n \in N\}$; $\{n \in I \text{ and } n > -9\}$; $\{2n + 1|n \in N \text{ and } n > 29\}$
    (b) Prove that any subset of a well ordered set is well ordered.
14. Show that if we assume Theorem 10 we may derive from it the principle of mathematical induction.
15. Let $N' = N \cup \{0, -1, -2, \ldots, -m\}$ and $S \subseteq N'$. Prove that if (1) $-m \in S$ and (2) $k \in S$ implies $k + 1 \in S$, then $S = N'$.

6. The Greatest Common Divisor

The simple fact that if we have two natural numbers a and b with $a > b$ we can divide a by b and obtain a non-negative remainder less than b is a matter of considerable significance. We generalize this fact slightly in our next theorem.

Theorem 13. The Division Algorithm. Given two integers a and b with $b > 0$, there exists a unique pair of integers q and r such that $a = bq + r$ where $0 \leq r < b$.

Let us first note that the algorithm can be made plausible by considering the multiples, bq, of b displayed on a line as shown below:

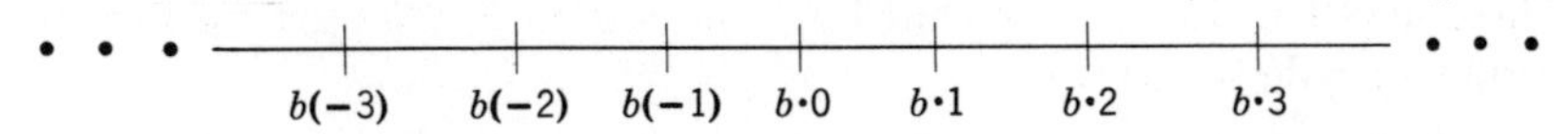

Clearly, the point representing a must fall in some one of the invervals determined by these points corresponding to the multiples bq. Suppose that it falls in the interval between bq and $b(q + 1)$ (exclusive of the right hand point). Then $a - bq = r$, where r represents a length shorter than the whole length of the interval and, hence, $0 \leq r < b$ as desired.

Now while the above argument makes Theorem 13 plausible, it is not a legitimate proof of the theorem since it rests on some unproved statements concerning the correspondence between numbers and points on a line. It does, however, motivate the proof which follows.

Consider the set $S = \{a - bx | x \in I \text{ and } a - bx \geq 0\}$. We note first that, since $b > 0$, it follows that $b \geq 1$ (Lemma 1) and hence $-|a|b \leq -|a| \leq a$. Now consider $y = a - (-|a|)b$. Then, since $-|a|b \leq a$, it follows that $y \geq 0$. Hence $y \in S$ and $S \neq \varnothing$. Thus either $0 \in S$, in which case there exists a $q \in S$ such that $a - bq = 0$, $a = bq + 0$; or else $0 \notin S$ but S contains a smallest positive integer, $r = a - bq$ (Theorem 10). Now $r > 0$ and if $r \geq b$, then $0 \leq r - b = a - bq - b = a - (q + 1)b < r$ contrary to the hypothesis that $0 \notin S$ and r is the smallest positive integer in S. Hence $r < b$ as desired.

To prove the uniqueness of q and r we assume a second pair of integers, q' and r', such that $a = bq' + r'$ where $0 \leq r' < b$. But then $bq' + r' = bq + r$ and $r' - r = (q - q')b$ so that $b|(r' - r)$. But $|r' - r| < b$ and hence $r' - r = 0$ and, since $b \neq 0$, $q' - q = 0$.

The integer r is called the *remainder* and the integer q the *quotient*.

The division algorithm, so basic in our further development, is capable of considerable generalization. It cannot, however, be generalized to an arbitrary integral domain even if the integral domain is ordered. Thus, for example, if we consider the integral domain R and the elements $\frac{3}{4}$ and $\frac{1}{2}$ in R we can write

$$\tfrac{3}{4} = \tfrac{1}{2} \cdot \tfrac{3}{2} + 0 \text{ where } 0 \leq 0 < \tfrac{1}{2}$$

$$\tfrac{3}{4} = \tfrac{1}{2} \cdot \tfrac{1}{1} + \tfrac{1}{4} \text{ where } 0 \leq \tfrac{1}{4} < \tfrac{1}{2}$$

etc. so that there is not a unique quotient or remainder satisfying the conditions of Theorem 13.

On the other hand, it is sometimes possible to develop a useful form of a division algorithm for integral domains that are not ordered. We shall do this in Chapter 12 for polynomials. In other situations a norm (Section 3) may be used. Thus in the integral domain $G = \{a + bi | a, b \in I\}$ we define $N(a + bi) = a^2 + b^2$ and may prove that for $\alpha, \beta \in G$ with $\beta \neq 0$, there exist $\gamma, \rho \in G$ such that $\alpha = \beta\gamma + \rho$ where $N(\rho) < N(\beta)$. We shall not pursue these questions further here but leave the matter as a subject for further reading for the student.

Definition 16. *We will call an integer d a* ***greatest common divisor*** *of two integers a and b* $(ab \neq 0)$ *if*
(1) $d|a$ *and* $d|b$
and
(2) *if* $c|a$ *and* $c|b$, *then* $c|d$.

The next theorem will prove that any two nonzero integers have a unique positive greatest common divisor. We will speak of this as *the* greatest common divisor or g.c.d. and write the g.c.d. of a and b as (a, b). If $(a, b) = 1$ we will say that a and b are *relatively prime*. Thus, for example, $(9, 12) = 3$, $(4, 20) = 4$, $(14, 9) = 1$.

Theorem 14. The Euclidean Algorithm. Any two nonzero integers a and b have a unique positive g.c.d.

Proof. We first observe that if a and b have both d and d' as a g.c.d. then d and d' are associates. For $d|d'$ and $d'|d$ by Definition 16 and hence there exist integers q and q' such that $d = d'q$ and $d' = dq'$. But then $d = (dq')q = d(q'q)$ and $q'q = 1$. Hence $q|1$ and d is an associate of d'. Since the only units in I are 1 and -1 we conclude that if two nonzero integers have a g.c.d. they have a positive g.c.d.

To find this positive g.c.d. we proceed as follows. Without loss of generality we may assume that $a > 0$ and $b > 0$. Now write $a = bq + r$, $0 \leq r < b$. If $r = 0$, $b = (a, b)$. If $r \neq 0$, $(a, b) = (b, r)$. For, suppose $d = (a, b)$ and $d' = (b, r)$. Then $d|a$ and $d|b$ so that d divides $a - bq = r$. Thus $d|b$ and $d|r$ and hence $d|d'$. On the other hand, $d'|b$ and $d'|r$ and thus d' divides $bq + r = a$. Hence $d'|a$ and $d'|b$ so that $d'|d$. Therefore $d = d'$.

We have seen that we can reduce the problem of finding (a, b) to the problem of finding (b, r). We continue by applying the division algorithm to b and r to obtain $b = rq_1 + r_1$, $0 \leq r_1 < r$. As before we conclude that if $r_1 = 0$, $r = (b, r)$ and if $r_1 \neq 0$, $(b, r) = (r, r_1)$. Continuing in this manner we obtain the following sequence of equalities and inequalities:

$$a = bq + r \qquad 0 < r < b$$

$$b = rq_1 + r_1 \qquad 0 < r_1 < r$$

$$r = r_1q_2 + r_2 \qquad 0 < r_2 < r_1$$

$$r_1 = r_2q_3 + r_3 \qquad 0 < r_3 < r_2$$

$$r_j = r_{j+1}q_{j+2} + r_{j+2} \qquad 0 < r_{j+2} < r_{j+1}$$

Since the r_j form a decreasing set of non-negative integers, there must exist a natural number n such that $r_{n+1} = 0$. Thus our sequence of equalities ends with

$$r_{n-2} = r_{n-1}q_n + r_n \qquad 0 < r_n < r_{n-1}$$

$$r_{n-1} = r_nq_{n+1}$$

But then $(a, b) = (b, r) = (r, r_1) = (r_1, r_2) = \ldots = (r_{n-1}, r_n) = r_n$.

Theorem 15. There exist integers m and n such that $(a, b) = ma + nb$.

Proof. From the above series of equalities in the proof of Theorem 14 we have

$$r_n = (a, b) = r_{n-2} - r_{n-1}q_n. \tag{1}$$

Then from the equality

$$r_{n-3} = r_{n-2}q_{n-1} + r_{n-1}$$

we have

$$r_{n-1} = r_{n-3} - r_{n-2}q_{n-1}. \tag{2}$$

Using (2) in (1) we obtain

$$\begin{aligned}(a, b) &= r_{n-2} - (r_{n-3} - r_{n-2}q_{n-1})q_n \\ &= (1 + q_{n-1}q_n)r_{n-2} - r_{n-3}q_n.\end{aligned} \tag{3}$$

Thus we have expressed (a, b) as a linear combination of r_{n-1} and r_{n-3}. We may now use the equality

$$r_{n-4} = r_{n-3}q_{n-2} + r_{n-2}$$

in (3) to obtain (a, b) as a linear combination of r_{n-3} and r_{n-4}. Continuing in this fashion we eventually arrive at (a, b) as a linear combination of r_1 and r. Then we replace r_1 by $b - rq_1$ and, finally, r by $a - bq$ to arrive at the desired linear combination of a and b.

Example. Find (432,342) and express it in the form $432m + 342n$.

Solution.

$$\begin{aligned}432 &= 1\cdot 342 + 90 \\ 342 &= 3\cdot 90 + 72 \\ 90 &= 1\cdot 72 + 18 \\ 72 &= 4\cdot 18\end{aligned}$$

Thus (432, 342) = 18. Now we have

$$\begin{aligned} 18 &= \underline{90} - 1\cdot\underline{72} \\ &= \underline{90} - 1(\underline{342} - 3\cdot\underline{90}) = 4\cdot\underline{90} - 1\cdot\underline{342} \\ &= 4(\underline{432} - 1\cdot\underline{342}) - 1\cdot\underline{342} = 4\cdot\underline{432} + (-5)(\underline{342}) \end{aligned}$$

where we have underlined the remainders and 432 and 342 to keep track of them.

Note that the m and n of Theorem 15 are not unique. For example,

$$18 = (4 + 342)\,432 - (5 + 432)\,342 = 346\cdot 432 + (-437)\,342.$$

Exercises 5.6

1. Find (595, 252) and express it in the form $595m + 252n$ in two ways.
2. Do as in problem 1 for (294, 273).
3. Do as in problem 1 for (163, 34).
4. Do as in problem 1 for (6432, 132).
5. Find (3456, 7234).
6. Prove that if $m > 0$, $(ma, mb) = m\,(a, b)$.
7. Prove that $((a, b), c) = (a, (b, c)) = ((a, c), b)$.
8. Prove that $(a, m) = (b, m) = 1$ if and only if $(ab, m) = 1$.
9. Prove that if there exist integral solutions, x, y, of the equation $ax + by = 1$ where a and b are integers, then $(a, b) = 1$.
10. Prove that if the equation $ax + by = d$, a and b integers, $(a, b) = d$, has a solution $x = m$ and $y = n$, then $x = m + bt$, $y = n + at$ is also a solution for every integer t.
***11.** Let $*$ be a binary operation on N such that (a) $a * b = b * a$, (b) $a * a = a$, and (c) $a * (a + b) = a * b$ for all $a, b \in N$. (1) Prove that $a * b = (a, b)$; (2) show that if any one of the conditions (a), (b), or (c) are omitted that $a * b$ need not be equal to (a, b).

7. Unique Factorization in *I*

Theorem 16. If p is a prime and $p|ab$ where $a, b \in I$, then $p|a$ or $p|b$.

Proof. Suppose $p\nmid a$. Since p is a prime, $(p, a) = 1$. By Theorem 15 there exist integers m and n such that $(p, a) = 1 = mp + na$ and hence $b = mpb + nab$. Since $p|ab$, p is a divisor of $mpb + nab$ and hence $p|b$.

Corollary. If p is a prime and $p \mid a_1a_2 \ldots a_n$ where $a_1, a_2, \ldots, a_n \in I$, then there exists an integer i, $1 \le i \le n$, such that $p|a_i$.

Proof. This follows by repeated application of Theorem 16. The details are left to the student.

We are now ready to prove the unique factorization theorem for integers.

Theorem 17. Every nonzero integer a (except 1 and -1) can be expressed as a unit of I times a product of positive primes. This representation is unique except for the order in which the prime factors occur.

Proof. Clearly we can confine our attention to positive integers since if $n > 0$ is a product of positive primes, $n = p_1p_2 \ldots p_k$, then $-n = (-1)p_1p_2 \ldots p_k$ is a product of positive primes times a unit of I. Now let S be the set of all positive integers for which the result is *not* true. If $S \neq \varnothing$, then there exists, by the well-ordering principle, a smallest positive integer m such that m is not a product of primes. Obviously, then, m is itself not a prime so that $m = m_1m_2$ where $1 < m_1 < m$ and $1 < m_2 < m$. But, by definition of m, both m_1 and m_2 are products of primes so that $m_1 = p_1p_2 \ldots p_k$, $m_2 = q_1q_2 \ldots q_s$ where the p_i and q_i are positive primes. Hence $m = p_1p_2 \ldots p_kq_1q_2 \ldots q_s$, a contradiction. Thus $S = \varnothing$.

Suppose now that, for some integer a, we have $|a| = p_1p_2 \ldots p_k$ (p_i positive primes) and also $|a| = q_1q_2 \ldots q_s$ (q_i positive primes). Then

$$p_1p_2 \ldots p_k = q_1q_2 \ldots q_s$$

Without loss of generality we may assume that $s \geq k$. By the corollary to Theorem 16, p_1 divides one of the q_i. Since we are not concerned with the ordering, we may assume that this q_i is q_1. But since p_1 and q_1 are both positive primes, $p_1 = q_1$, and hence

$$p_2p_3 \ldots p_k = q_2q_3 \ldots q_s$$

Continuing, we get $p_2 = q_2, q_3 = p_3, \ldots, p_k = q_k$. Then, if $s \neq k$, we have $q_{k+1}q_{k+2} \ldots q_s = 1$, a contradiction since $q_{k+1} \nmid 1$. Hence $s = k$.

We note that the theorem does not exclude the occurrence of equal primes. Hence we may write.

$$a = \pm\, p_1^{\alpha_1}p_2^{\alpha_2} \ldots p_t^{\alpha_t}$$

and have proved that the exponents $\alpha_1, \alpha_2, \ldots, \alpha_t$ as well as the primes $p_1, p_2, \ldots, p_t$ are uniquely determined.

Let us return now to the topic of residue classes to prove the following theorem.

Theorem 18. J_m is an integral domain if and only if m is a prime.

Proof. It is obvious that the condition that m be a prime is necessary since if $m = m_1m_2$, $1 < m_1 < m$, $1 < m_2 < m$, then $C_{m_1}C_{m_2} = C_0$ as, for example, in J_6 we have $C_2C_3 = C_0$. Furthermore, if $m = 1$, we have only one element, C_0, contrary to our requirement that an integral domain have at least two elements.

Whether or not m is a prime, properties 1, 2, and 5 of Definition 1 follow easily from Definition 3 and the properties of integers. For property 3 we observe that $C_i + C_0 = C_i$; for property 4 that $C_i + C_{m-i} = C_0$ and

$C_0 + C_0 = C_0$†; and for property 6 that $C_1C_j = C_jC_1 = C_j$. Now for property 7 we observe that if $C_iC_j = C_0$, then $0 \equiv ij \pmod{m}$. Hence $m|ij$ so that if m is a prime, $m|i$ or $m|j$ by Theorem 16. But $i, j < m$ and hence $i = 0$ or $j = 0$.

Exercises 5.7

1. Prove that if $d = (a, b)$, $a = a_1d$, and $b = b_1d$, then $(a_1, b_1) = 1$.
2. Prove that if $(a, b) = 1$, $a|c$, and $b|c$, then $ab|c$.
3. Prove that if $(a, c) = 1$ and $c|ab$, then $c|b$.
4. Prove the corollary to Theorem 16.
5. Prove that if $(c, d) = 1$, then $(c^n, d) = 1$ for all $n \in N$.
6. Prove that if $(a, b) = 1$, then $(a + b, a - b) = 1$ or 2.
7. If $(a, c) \equiv d$, prove that $a|b$ and $c|d$ if and only if $ac|bd$.

*8. Prove that the number of primes is infinite. (*Hint*: Assume that the number of primes is finite and form their product $p_1p_2 \ldots p_n$. Now consider $p_1p_2 \ldots p_n + 1$.)

PROBLEMS FOR FURTHER INVESTIGATION

1. Investigate the solution of equations in J_m. For example, does the linear equation $ax + b = 0$. $(a, b \in J_m)$ always have a solution in J_m? If it does have a solution, is it unique? What about quadratic equations, higher degree equations, systems of simultaneous linear equations? [N3]
2. A function, f, on an integral domain D is defined as a correspondence which associates with each $x \in D$ an element $f(x) \in D$. Thus, for example, we have the function f which associates with x its square, x^2, and we write $f(x) = x^2$. Can you form definitions of equality, addition, and multiplication of functions so as to get a ring of functions? For arbitrary functions will the ring be commutative? Can you restrict the functions so that the ring will be commutative? [B4]
3. Investigate factorization in $\{a + bi | a, b \in I\}$, the integral domain of *Gaussian integers*. (For a complete development you will need to develop a Euclidean algorithm for Gaussian integers.) [B4], [P1]
4. Define the least common multiple, $[a, b]$, of two integers, a and b, in a manner similar to our definition of the g.c.d. Develop properties of $[a, b]$ and relate $[a, b]$ to (a, b). [O1]
5. Let $V_p(a)$ denote the highest power of the prime p dividing the nonzero integer a and define $||a|| = 2^{-V_p(a)}$. Compare the properties of this variation of absolute value with ordinary absolute value. (For example, consider $||a + b||$.) [B4]
6. Find a method of determining all possible solutions in integers of the equation $x^2 + y^2 = z^2$. [N3], [O1]

REFERENCES FOR FURTHER READING

[B4], [P1].

† In situations like this, a common error is to assume that the additive inverse to an element, x, is simply $-x$ and thus, in this case to write $-C_i$ for the additive inverse of C_i. But we must show that $-C_i = C_j$ for some $j = 0, 1, \ldots, m - 1$.

6

The Rational Numbers

Our treatment of the rational numbers as classes of pairs of integers (a, b) $(b \neq 0)$ will so closely parallel our treatment of the integers as classes of pairs of natural numbers that we will only sketch the procedure here and leave the details to be filled in by the student.

Our motivation is based on the solution of equations of the form $bx = a$ (a and b integers, $b \neq 0$). If, for example, $b = 2$ and $a = -6$ we have a solution, -3, in integers. In general, if $b|a$ we see that a and b, together with the equation $bx = a$, determine an integer, a/b, which is the (unique) solution to the equation: that is $\{x|bx = a\} = \{a/b\}$.

On the other hand, if $b = 5$ and $a = 2$, the equation $bx = a$ has no solution in integers but we can still consider the ordered pair $(2, 5)$ analogously to the ordered pair $(-6, 2)$ in our first example.†

1. Equivalence

Let a, b, c, d, and x be integers with $bd \neq 0$ and such that $bx = a$ and $dx = c$. Then our properties of integers assure us that $d(bx) = b(dx) = da$. Since $dx = c$ we have $bc = da = ad$. Thus we see that if the ordered pairs (a, b) and (c, d) both define the same integer (as, for example, do $(6, 3)$ and $(8, 4)$) we must have $bc = ad$ $(3 \cdot 8 = 6 \cdot 4$ in our example). These considerations lead us to make the following definition of equivalence for any two ordered pairs of integers (a, b) and (c, d) with $bd \neq 0$.

Definition 1. *If $bd \neq 0$ then, $(a, b) \sim (c, d)$ (read "(a, b) equivalent to (c, d)") if and only if $ad = bc$.*

We let

$$S = \{(a, b)|a, b \in I, b \neq 0\}$$

† The notation (a, b) gets a heavy workout having been used before in connection with integers and also as a symbol for the g.c.d. We could, of course, use another notation such as $[a, b]$ or $\langle a, b \rangle$ but, in practice, the meaning of (a, b) will always be clear from the context. A similar remark applies to our use of "$\sim$."

Theorem 1. The relation of $\sim$ in Definition 1 is an equivalence relation on S.

Proof. (1) $(a, b) \sim (a, b)$ since $ab = ba$.

(2) $(a, b) \sim (c, d)$ means that $ad = bc$. Hence $cb = da$ and thus $(c, d) \sim (a, b)$.

(3) If $(a, b) \sim (c, d)$ and $(c, d) \sim (e, f)$, then $ad = bc$ and $cf = de$. Hence $e(ad) = e(bc)$ so $a(de) = e(bc)$. Since $de = cf$ we then have $a(cf) = e(bc)$ so $(af)c = (be)c$. If $c \neq 0$ the cancellation law of multiplication for integers gives us $af = be$ so that $(a, b) \sim (e, f)$ as desired. On the other hand, if $c = 0$ we have $ad = bc = 0$ and, since $d \neq 0$, $a = 0$. Similarly $de = cf = 0$ and hence $e = 0$. Thus $cf = 0 = de$ and, again, $(a, b) \sim (e, f)$.

By Theorem 4.1, our equivalence relation partitions the set S. We now define a rational number in terms of these equivalence classes.

Definition 2. *If* $b \neq 0$, *the rational number* a/b (*or* $\frac{a}{b}$) *is defined as* $R\langle a, b\rangle = \{(x, y) | x, y \in I$ *and* $(x, y) \sim (a, b)\}$.

Example. $2/3 = R\langle 2, 3\rangle = \{\ldots, (-4, -6), (-2, -3), (2, 3), (4, 6, \ldots)\}$.

2. Addition and Multiplication of Rational Numbers

Let us first motivate our definitions as follows: We suppose a, b, c, d, x, and y are all integers with $bd \neq 0$ and that $bx = a$ and $dy = c$. Then by the properties of integers, we have

$$d(bx) + b(dy) = (bd)x + (bd)y = (bd)(x + y) = da + bc = ad + bc.$$

That is, if x and y correspond to the ordered pairs (a, b) and (c, d) respectively, $x + y$ corresponds to the ordered pair $(ad + bc, bd)$.

Similarly, $(bx)(dy) = (bd)(xy) = ac$ and hence xy corresponds to the ordered pair (ac, bd).

Definition 3. (1) $R\langle a, b\rangle + R\langle c, d\rangle = R\langle ad + bc, bd\rangle$.
(2) $R\langle a, b\rangle \times R\langle c, d\rangle = R\langle ac, bd\rangle$.

Theorem 2. If $R\langle a, b\rangle = R\langle a', b'\rangle$ and $R\langle c, d\rangle = R\langle c', d'\rangle$, then $R\langle a, b\rangle + R\langle c, d\rangle = R\langle a', b'\rangle + R\langle c', d'\rangle$ and $R\langle a, b\rangle \times R\langle c, d\rangle = R\langle a', b'\rangle \times R\langle c', d'\rangle$.

Proof. $R\langle a, b\rangle + R\langle c, d\rangle = R\langle ad + bc, bd\rangle = \{(x, y) | x, y \in I$ and $(x, y) \sim (ad + bc, bd)\}$ and $R\langle a', b'\rangle + R\langle c', d'\rangle = R\langle a'd' + b'c', b'd'\rangle = \{(x, y) | x, y \in I$ and $(x, y) \sim (a'd' + b'c', b'd')\}$.

Clearly, $R\langle ad + bc, bd\rangle = R\langle a'd' + b'c', b'd'\rangle$ if and only if $(ad + bc, bd) \sim (a'd' + b'c', b'd')$ and since $R\langle a, b\rangle = R\langle a', b'\rangle$ and $R\langle c, d\rangle = R\langle c', d'\rangle$ we have $(a, b) \sim (a', b')$ and $(c, d) \sim (c', d')$. Thus $ab' = ba'$ and

$cd' = dc'$. Hence

$$(ab')(dd') = (ba')(dd') \text{ and } (cd')(bb') = (dc')(bb')$$

Thus

$$(ab')(dd') + (cd')(bb') = (ba')(dd') + (dc')(bb')$$

so that

$$(ad + bc)(b'd') = (bd)(a'd' + b'c')$$

and

$$(ad + bc, bd) \sim (a'd' + b'c', b'd');$$

$$R\langle ad + bc, bd)\rangle = R\langle a'd' + b'c', b'd'\rangle$$

The proof that $R\langle a, b\rangle \times R\langle c, d\rangle = R\langle a', b'\rangle \times R\langle c', d'\rangle$ is similar and is left as an exercise for the student.

Exercises 6.2

1. Perform the following computations and write your answers in the form $R\langle a, b\rangle$ with a and b relatively prime.
(a) $R\langle 3, 4\rangle + R\langle -1, 2\rangle$ (b) $R\langle 5, 12\rangle + R\langle 3, -8\rangle$
(c) $R\langle 0, 1\rangle + R\langle -7, 9\rangle$ (d) $R\langle -3, 4\rangle \times R\langle 2, -5\rangle$
(e) $R\langle 6, 5\rangle \times R\langle -3, -4\rangle$ (f) $R\langle 1, 1\rangle \times R\langle -7, 9\rangle$

2. Verify Theorem 2 for the following cases:
(a) $R\langle 2, 3\rangle = R\langle 4, 6\rangle, R\langle -5, 2\rangle = R\langle 15, -6\rangle$
(b) $R\langle -1, -1\rangle = R\langle 3, 3\rangle, R\langle 1, 7\rangle = R\langle 2, 14\rangle$
(c) $R\langle 0, 4\rangle = R\langle 0, -1\rangle, R\langle -3, 4\rangle = R\langle -6, 8\rangle$

3. Complete the proof of Theorem 2.

3. Properties of the Rational Numbers

It is convenient to define

$$(a, b) \oplus (c, d) = (ad + bc, bd) \quad \text{and} \quad (a, b) \otimes (c, d) = (ac, bd)$$

and to observe that

$$R\langle a, b\rangle + R\langle c, d\rangle = R\langle (a, b) \oplus (c, d)\rangle \text{ and}$$

$$R\langle a, b\rangle \times R\langle c, d\rangle = R\langle (a, b) \otimes (c, d)\rangle$$

Theorem 3. The operations of addition and multiplication on the rational numbers are commutative and associative, and multiplication is distributive with respect to addition.

Proof. We prove only the commutative law of addition and the associative law of multiplication, leaving the others as exercises for the student.

From our previous remarks it follows that we need only to show that

(1) $(a, b) \oplus (c, d) = (c, d) \oplus (a, b)$

and

(2) $[(a, b) \otimes (c, d)] \otimes (s, t) = (a, b) \otimes [(c, d) \otimes (s, t)]$.

To prove (1) we observe that

$(a, b) \oplus (c, d) = (ad + bc, bd)$	Definition of $\oplus$
$= (da + cb, db)$	Commutative law of multiplication in I
$= (cb + da, db)$	Commutative law of addition in I
$= (c, d) \oplus (a, b)$	Definition of $\oplus$

To prove (2) we observe that

$[(a, b) \otimes (c, d)] \otimes (s, t)$

$= (ac, bd) \otimes (s, t)$	Definition of $\otimes$
$= ([ac]s, [bd]t)$	Definition of $\otimes$
$= (a[cs], b[dt])$	Associative law of multiplication in I
$= (a, b) \otimes (cs, dt)$	Definition of $\otimes$
$= (a, b) \otimes [(c, d) \otimes (s, t)]$	Definition of $\otimes$.

Theorem 4. Cancellation laws for rational numbers.

(1) If $R\langle a, b\rangle + R\langle c, d\rangle = R\langle e, f\rangle + R\langle c, d\rangle$, then $R\langle a, b\rangle = R\langle e, f\rangle$

(2) If $R\langle a, b\rangle \times R\langle c, d\rangle = R\langle e, f\rangle \times R\langle c, d\rangle$ and $c \neq 0$, then $R\langle a, b\rangle = R\langle e, f\rangle$.

Proof. Again, we "translate" these properties to

(1) If $(a, b) \oplus (c, d) \sim (e, f) \oplus (c, d)$, then $(a, b) \sim (e, f)$

and

(2) If $(a, b) \otimes (c, d) \sim (e, f) \otimes (c, d)$ and $c \neq 0$, then $(a, b) \sim (e, f)$.

To prove (1) we observe that $(a, b) \oplus (c, d) = (ad + bc, bd)$ and $(e, f) \oplus (c, d) = (ed + fc, fd)$. Hence we have $(ad + bc, bd) \sim (ed + fc, fd)$ or

$$(ad + bc)(fd) = (bd)(ed + fc)$$

Continuing, we have

$(ad + bc)(fd) = (ed + fc)(bd)$	Commutative law of multiplication in I
$[(ad + bc)f]d = [(ed + fc)b]d$	Associative law of multiplication in I
$(ad + bc)f = (ed + fc)b$	Cancellation law for multiplication in I $(d \neq 0)$
$(ad)f + (bc)f = (ed)b + (fc)b$	Distributive law in I
$(ad)f + f(cb) = (ed)b + (fc)b$	Commutative law of multiplication in I
$(ad)f + (fc)b = (ed)b + (fc)b$	Associative law of multiplication in I
$(ad)f = (ed)b$	Cancellation law for addition in I
$a(df) = e(db)$	Associative law of multiplication in I
$a(fd) = e(bd)$	Commutative law of multiplication in I
$(af)d = (eb)d$	Associative law of multiplication in I
$af = eb$	Cancellation law for multiplication in I $(d \neq 0)$
$af = be$	Commutative law of multiplication in I
$(a, b) \sim (e, f)$	Definition of $\sim$

Note. We have given a complete sequence of steps here. Once the logical necessity for such details is realized, however, it is certainly in order to proceed more quickly. Thus we might say that, by properties of the integers, $(ad)f = (ed)b$ implies $af = be$ since $d \neq 0$. It is very important, however, that the student be able to give such detailed proofs and until he is certain that he can do so he should avoid combining steps.

The proof of the cancellation law for multiplication is similar and is left as an exercise for the student.

We let R be the set of all rational numbers, that is, $R = \{R\langle x, y\rangle | x, y \in I, y \neq 0\}$.

Theorem 5. There is a unique multiplicative identity, $R\langle 1, 1\rangle$, in R and a unique additive identity, $R\langle 0, 1\rangle$.

Proof. $R\langle 1, 1\rangle \times R\langle a, b\rangle = R\langle 1 \times a, 1 \times b\rangle = R\langle a, b\rangle$. If $R\langle r, s\rangle \times R\langle a, b\rangle = R\langle a, b\rangle$ we have $R\langle r, s\rangle \times R\langle a, b\rangle = R\langle 1, 1\rangle \times R\langle a, b\rangle$ for *all* rational numbers $R\langle a, b\rangle$. Choosing a rational number $R\langle a, b\rangle$ with $a \neq 0$ we may apply the cancellation law for multiplication for rational numbers to obtain $R\langle r, s\rangle = R\langle 1,1 \rangle$.

We leave the proof of the existence of a unique additive identity as an exercise for the student.

Theorem 6. If $R\langle a, b\rangle$ and $R\langle c, d\rangle$ are any two rational numbers, there exists a unique rational number, $R\langle x, y\rangle$, such that

$$R\langle a, b\rangle + R\langle x, y\rangle = R\langle c, d\rangle$$

We have $R\langle x, y\rangle = R\langle cb + d(-a), db\rangle$.

Proof. $R\langle a, b\rangle + R\langle cb + d(-a), db\rangle = R\langle a(db) + b[cb + d(-a)], b(db)\rangle$. By a liberal use of the properties of the integers† we have

$$a(db) + b[cb + d(-a)] = b^2c$$

where we use the fact that $b[d(-a)] = -[a(db)]$. Thus

$$R\langle a(db) + b[cb + d(-a)], b(db)\rangle = R\langle b^2c, b^2d\rangle$$

We leave as exercises for the student the proofs that $R\langle b^2c, b^2d\rangle = R\langle c, d\rangle$ and that $R\langle x, y\rangle$ is unique.

The student should note that $R\langle cb + d(-a), db\rangle$ is simply our formal version of $c/d - a/b = (cb - da)/db$ and that, in particular, the equation $R\langle a, b\rangle + R\langle x, y\rangle = R\langle 0, 1\rangle$ has a unique solution. Hence every rational number has a unique additive inverse.

Exercises 6.3

1. Prove the commutative law of multiplication for rational numbers.
2. Prove the associative law of addition for rational numbers.
3. Prove the distributive law for multiplication with respect to addition for rational numbers.
4. Prove the cancellation law for multiplication of rational numbers.
5. Prove that $R\langle 0, 1 \rangle$ is a unique additive identity for the rational numbers.
6. Prove that if m is a nonzero integer, $R\langle ma, mb \rangle = R\langle a, b \rangle$.
7. Prove that, in Theorem 6, $R\langle x, y \rangle$ is unique.
8. Prove that the set, R, of all rational numbers is an integral domain.

4. The Integers as a Subset of the Rational Numbers

We recall that $a/b = R\langle a, b\rangle$ (Definition 2).

† Recall the remark following the proof of Theorem 4.

Theorem 7. The subset $I' = \{a/1 | a \in I\}$ of R is isomorphic to I under the correspondence $a \in I \leftrightarrow a/1 \in I'$.

Proof. It is obvious that the correspondence is 1–1. Now $a \leftrightarrow a/1$, $b \leftrightarrow b/1$, and $a + b \leftrightarrow (a + b)/1$. Since $a/1 + b/1 = (a \cdot 1 + 1 \cdot b)/1 \cdot 1 = (a + b)/1$ we have $a + b \leftrightarrow a/1 + b/1$ as desired. Similarly, $ab \leftrightarrow (ab)/1 = (a/1) \times (b/1)$.

From now on we will identify a with $a/1$ even though we only have $a \leftrightarrow a/1$. (Cf. $(a + 1, 1)$ and a for integers.)

5. Additive and Multiplicative Inverses in *R*

In Theorem 5 we noted that $R\langle 0, 1\rangle = 0/1$ is an additive identity for the rational numbers.

Definition 4. $-(a/b) = (-a)/b$.

Theorem 8. $-(a/b)$ is the unique additive inverse of a/b.

Proof. This follows from Theorem 6 with $c = 0$ and $d = 1$ since then $R\langle cb + d(-a), db\rangle = R\langle -a, b\rangle = (-a)/b = -(a/b)$.

Definition 5. $a/b - c/d = a/b + [-(c/d)]$.

In the following theorem we collect some useful facts about rational numbers.

Theorem 9. (1) $(-a)/b = a/(-b)$;
(2) $(-a)/(-b) = a/b$;
(3) $-[(-a)/b] = a/b$;
(4) $-[-(a/b)] = a/b$;
(5) $-(a/b - c/d) = -(a/b) + c/d$;
(6) $[-(a/b)][-(c/d)] = (a/b)(c/d)$.

Proof. We prove here (1) and (5) leaving the rest as exercises for the student. For (1) we observe that $(-a)/b = a/(-b)$ if and only if $(-a)(-b) = ba$. By the properties of integers, $(-a)(-b) = ab = ba$.

For (5) we observe that

$$-\left(\frac{a}{b} - \frac{c}{d}\right) = -\left[\frac{a}{b} + \left(-\frac{c}{d}\right)\right]$$

$$= -\left[\frac{a}{b} + \frac{(-c)}{d}\right]$$

$$= -\left[\frac{ad + b(-c)}{bd}\right]$$

$$= \frac{-[ad + b(-c)]}{bd}$$

$$= \frac{-(ad) - [b(-c)]}{bd}$$

$$= \frac{bc - ad}{bd}$$

by the definitions for rational numbers and the properties of the integers. On the other hand

$$-\frac{a}{b} + \frac{c}{d} = \frac{(-a)}{b} + \frac{c}{d}$$

$$= \frac{(-a)d + bc}{bd} = \frac{bc - ad}{bd}$$

Note that (5) can also be established as a consequence of the fact that R is an integral domain (problem 8 of Exercises 6.3) and the fact that if x and y are elements of an integral domain, $-(x - y) = -x + y$ (problem 4d of Exercises 5.2).

We pointed out in Theorem 5 that $R\langle 1, 1\rangle = 1/1$ is the multiplicative identity for the rational numbers. We now establish the existence of multiplicative inverses in R.

Theorem 10. If $a \neq 0$, the unique multiplicative inverse of a/b is b/a.

Proof. $(a/b)(b/a) = ab/ba = 1/1$ with uniqueness following from the cancellation law of multiplication for rational numbers.

Exercises 6.5

1. Prove that $(-a)/(-b) = a/b$.
2. Prove that $-[(-a)/b] = a/b$.
3. Prove that $-[-(a/b)] = a/b$.
4. Prove that $[-(a/b)][-(c/d)] = (a/b)(c/d)$.

6. The Ordering of the Rational Numbers.

Our next theorem has to do with the extension (Definition 5.14) of the unique (Theorem 5.6) ordering of I to an ordering of R and states that the extension is both possible and unique.

Theorem 11. The only ordering of R is that defined by $a/b > 0$ if and only if ab is a positive integer ($a, b \in I$, $b \neq 0$).

Proof. We may first observe that if the relation $a/b > 0$ defined by $ab > 0$ is an ordering, it is an extension of the ordering of I since, if $a \in I$, $a = a/1 > 0$ if and only if $a \cdot 1 = a > 0$. (Here, of course, we should really write $a/1 >_R 0$ and $a \cdot 1 >_P 0$ to indicate that, in the first case, we are talking about our newly defined ordering in R and, in the second case, the previously defined ordering in I. Which ordering we are considering is, however, always clear from the context.)

Next we show that if R has any ordering at all it must be the one given. For, given any $b \neq 0 \in I$, we have $b > 0$ or $-b > 0$ by condition (3) of Definition 5.11 and hence, in either case, $b^2 = b \cdot b = (-b)(-b) > 0$ by condition (2) of Definition 5.11. Thus if $a/b > 0$ we must have $(a/b)b^2 = ab > 0$.

It remains to show that the condition $a/b > 0$ if and only if $ab > 0$ actually defines an ordering. That is, by Definitions 5.11 and 5.12, we must show that:

(1) If $a/b > 0$ and $c/d > 0$, then $(a/b) + (c/d) > 0$;
(2) If $a/b > 0$ and $c/d > 0$, then $(a/b)(c/d) > 0$;
(3) One and only one of the alternatives $a/b > 0$, $-(a/b) > 0$, and $a/b = 0$ holds.

To prove (1) we observe that we have $ab > 0$, $cd > 0$, and

$$\frac{a}{b} + \frac{c}{d} = \frac{ad + bc}{bd}$$

Then $(a/b) + (c/d) > 0$ if and only if $(ad + bc)(bd) = d^2(ab) + b^2(cd) > 0$. Since d^2, b^2, ab, and cd are all positive in I we have our desired result.

Similarly, for (2) we have $(a/b)(c/d) = ac/bd$ and $ac/bd > 0$ if and only if $(ac)(bd) = (ab)(cd) > 0$. But $(ab)(cd) > 0$ since $ab > 0$ and $cd > 0$ in I.

For (3) we know that one and only one of the alternatives $ab > 0$, $-(ab) > 0$, and $ab = 0$ holds. If $ab = 0$ we have $a = 0$ since $b \neq 0$ and hence $a/b = 0$. If $ab > 0$ then $a/b > 0$ while if $-(ab) = (-a)b > 0$ we have $(-a)/b = -(a/b) > 0$.

One further point needs to be considered since there is nothing in our definition of $a/b > 0$ which says that we could not have, for example, $2/3 > 0$ but $4/6 < 0$ even though $2/3 = 4/6$. Now it is easy to check that $4/6 > 0$ but we now ask, in general: if $a/b > 0$ and $c/d = a/b$, is $c/d > 0$? To see that we have an affirmative answer we simply observe that $c/d = a/b$ means $ad = bc$. Then $(bc)(ad) = (bc)^2 > 0$ and since $ab > 0$ it follows that $cd > 0$.

Definition 6. *$a/b > c/d$ if and only if $(a/b) - (c/d) > 0$.*

Theorem 12. $a/b > c/d$ if and only if $ad > bc$.

Proof. By Definition 6, $a/b > c/d$ if and only if $(a/b) - (c/d) > 0$. Now $a/b - c/d = (ad - bc)/bd$. Hence, by Theorem 11, $a/b > c/d$ if and only if $(ad - bc)(bd) > 0$. Now by Theorem 9, $a/(-b) = (-a)/b$, so that we know that we may assume $b > 0$ and $d > 0$ and have $bd > 0$. Hence by the properties of ordering for the integers we conclude that $a/b > c/d$ if and only if $ad - bc > 0$ or $ad > bc$.

Theorem 13. (The Archimedean law) Let $x = a/b$ and $y = c/d$ be positive rational numbers. Then there exists $n \in N$ such that $nx > y$.

Proof. We know by Theorems 9 and 11 that we may assume a, b, c, and d to be positive integers. We wish to find $n \in N$ so that

$$nx = \frac{n}{1}\frac{a}{b} = \frac{na}{b} > \frac{c}{d}$$

By Theorem 12 this will happen if and only if

(1) $(na)d = n(ad) > bc$.

Now if $ad > bc$ we take $n = 1$ and if $ad = bc$ we take $n = 2$ for the desired result. If $ad < bc$ we use the division algorithm to obtain

$$bc = q(ad) + r \qquad (0 \leq r < ad)$$

Thus we have $ad > r$ so that

$$q(ad) + ad > q(ad) + r = bc$$

Hence

$$(q + 1)(ad) > bc$$

and (1) is satisfied by taking $n = q + 1$.

Exercises 6.6

If a and b are integers prove that

1. $1/a > 0$ if and only if $a > 0$.

2. $b > a > 0$ implies that $1/a > 1/b > 0$.

3. $0 > b > a$ implies that $0 > 1/a > 1/b$.

***4.** $a/b + a/c = a/(b + c)$ implies $a = 0$.

PROBLEMS FOR FURTHER INVESTIGATION

1. Outline a procedure for first constructing the positive rational numbers from the natural numbers and then constructing all of the rational numbers from the positive rational numbers. [R1]

2. We have been constructing new systems of numbers by considering ordered pairs of numbers from our previous system. Furthermore, we have motivated our discussion by considering the need for new numbers as indicated by equations. Consider now the equation $x^2 - 2 = 0$. Show that it has no solution in rational numbers and consider the possibility of using pairs of rational numbers to obtain a solution.

REFERENCES FOR FURTHER READING

[E2], [H3], [R1].

7

Groups and Fields

1. Definitions and Examples

In all of the systems we have studied so far (sets, natural numbers, integers, rational numbers, integral domains, and rings) we have always considered two operations on the system. If we concentrate on one operation we are led to the concept of a group in the sense of the following definition.

Definition 1. *A* ***group****, G, is a set of elements on which is defined a binary operation,* o, *such that*
(1) *There exists an element $e \in G$ such that $a \circ e = a$ for all $a \in G$ (existence of right hand identity);*
(2) *If $a \in G$, there exists an element $a' \in G$ such that $a \circ a' = e$ (existence of right hand inverse)†;*
(3) *If a, b, and $c \in G$, then $(a \circ b) \circ c = a \circ (b \circ c)$ (associative law).*

If a group consists of a finite number, n, of elements we call n the ***order*** *of the group.*

Example 1. The set of integers forms a group under the operation of addition. Here 0 is the identity and $-a$ the inverse of a. On the other hand, the set of non-negative integers does not form a group under addition since inverses do not exist for $a \neq 0$.

We may note that we have used here the "neutral" symbol, o, for our group operation. When, however, we are considering groups whose elements are numbers we normally use the ordinary signs of $+$ and $\times$ when our operation is addition or multiplication respectively. Also, it is frequently convenient to adopt the convention that ab shall mean $a \circ b$.

Example 2. The rational numbers form a group under addition and the nonzero rational numbers form a group under multiplication.

† As far as our postulates go there might be more than one right hand identity. Thus when we write $a \circ a' = e$ we should really say that a' is a right hand inverse of a *relative* to the right hand identity e. Later, however, we will show that there is only one right hand identity so that we omit this qualification.

Example 3. J_3 is a group of order 3 with respect to addition. C_0 is the identity for addition; C_0 is the additive inverse of C_0; C_1 is the additive inverse of C_2; and C_2 is the additive inverse of C_1. Note that this is a *finite* group whereas the groups of Examples 1 and 2 are *infinite* groups.

Example 4. J_3 with C_0 omitted is a group of order 2 with respect to multiplication. C_1 is the identity for multiplication; C_1 is the multiplicative inverse of C_1 and C_2 is the multiplicative inverse of C_2. On the other hand, J_4 with C_0 omitted is not a group with respect to multiplication. C_1 is the multiplicative identity but $C_2 \cdot C_2 = C_0$ so that we do not have closure.

We note in all of these examples that $a \circ b = b \circ a$ for all a and b in the group although this is not a requirement given in Definition 1. In the next section we will give some examples of groups where $a \circ b \neq b \circ a$ but right now we will go on to a consideration of fields. To do this it is first convenient to give a name to groups where $a \circ b = b \circ a$ for all a and b in the group.

Definition 2. *A group G is said to be* ***commutative*** *or* ***Abelian***† *if for all $a, b \in G$ it is true that $a \circ b = b \circ a$.*

If we compare the definition of an Abelian group with the definition of an integral domain, we see that we may describe an integral domain as a set D which possesses two binary operations, addition $(+)$, and multiplication $(\times)$, such that

1. D is an Abelian group under addition;
2. Multiplication is commutative and associative, and is distributive with respect to addition;
3. There exists a multiplicative identity;
4. D has at least two elements;
5. $a \times b = 0$ implies $a = 0$ or $b = 0$ (where 0 is the additive identity).

Now we note that were it not for the possible absence of multiplicative inverses of nonzero elements, D with 0 omitted would also be a group with respect to multiplication. Indeed, in some integral domains, the nonzero elements do have inverses as is the case for the integral domain of rational numbers. We are thus led to the definition of a field.

Definition 3. *A* ***field*** *is an integral domain in which every element except the additive identity has a multiplicative inverse.*

It is clear that the rational numbers form a field since, when $a \neq 0$, $(a/b)(b/a) = 1$. It is also true, as we will see, that the real numbers form a field and that the complex numbers form a field. We now give two additional examples of fields.

† In honor of Niels Henrik Abel (1802–1829), one of the founders of group theory.

Example 5. Let $F = \{a + b\sqrt{2} | a, b \in R\}$. Then F is a field under the usual operations of addition and multiplication. For we have noted before that F is an integral domain (problem 4, Exercises 5.1) and hence we need only to show that, if $a + b\sqrt{2} \neq 0$, $a + b\sqrt{2}$ has a multiplicative inverse.†

To do this we observe that

$$(a + b\sqrt{2})(a - b\sqrt{2}) = a^2 - 2b^2$$

and that, if $a + b\sqrt{2} \neq 0$, $a^2 - 2b^2 \neq 0$. For if $a^2 - 2b^2 = 0$, then $a^2/b^2 = (a/b)^2 = 2$ so that $\sqrt{2}$ would be a rational number. But (see Chapter 8), $\sqrt{2}$ is not a rational number and hence $a^2 - 2b^2 \neq 0$. Then if $c = a/(a^2 - 2b^2)$ and $d = -b/(a^2 - 2b^2)$, we have

$$(a + b\sqrt{2})(c + d\sqrt{2}) = 1.$$

(Cf. with the usual process of "rationalizing the denominator" of $1/(a + b\sqrt{2})$).

Example 6. In J_3 we have $C_1C_1 = C_1$ and $C_2C_2 = C_1$ so that J_3 is a field. More generally, we may show that J_p is a field for any prime p.

Theorem 1. J_t is a field if and only if t is a prime.

Proof. First, if t is not a prime, then $t = rs$ where $0 < r < t$ and $0 < s < t$. Thus $C_rC_s = C_0$. Now if C_r had a multiplicative inverse, C_k, we would have $C_s = C_1C_s = (C_kC_r)C_s = C_k(C_rC_s) = C_kC_0 = C_0$, a contradiction. Hence it is necessary that t be a prime.

Now suppose that t is a prime and consider any C_k where $0 < k < t$. We have $(t, k) = 1$ so that, by Theorem 5.15, there exist integers m and $n \neq 0$ such that $mt + nk = 1$. But this means that $nk \equiv 1 \pmod{t}$. Now for any integer $n \neq 0$ we know that there exists an integer n' such that $0 < n' < t$ and $n \equiv n' \pmod{t}$. (It is, of course, possible that $n' = n$.) For this n' we have, by the properties of congruences, $n'k \equiv 1 \pmod{t}$. But then $C_{n'}C_k = C_kC_{n'} = C_1$ and thus C_k has the multiplicative inverse $C_{n'}$.

Exercises 7.1

1. Do the even integers form a group with respect to addition? Do the odd integers?
2. Do the positive irrational numbers form a group with respect to multiplication?
3. In J_7 which of the following sets form a group with respect to multiplication: (a) $\{C_1, C_2, C_4\}$; (b) $\{C_0, C_1, C_2, C_3, C_4, C_5, C_6)$; (c) $\{C_1, C_6\}$; (d) $\{C_1, C_3, C_4, C_5\}$; (e) $\{C_1, C_2, C_3, C_4, C_5, C_6\}$?

† It is *not* sufficient just to write $1/(a + b\sqrt{2})$. Rather, we must explicitly exhibit a number of the form $c + d\sqrt{2}$ where $c, d \in R$ such that $(a + b\sqrt{2})(c + d\sqrt{2}) = 1$.

4. Let $G = \{C_k | C_k \in J_m,\ k \neq 0, \text{ and } (k, m) = 1\}$. Prove that G is a group under multiplication.
5. Which of the integral domains of problems 1–7 of Exercises 5.1 are fields?
*6. Let P be the set of all positive real numbers and q a fixed positive real number with $q \neq 1$. For $a, b \in P$ define

$$a \oplus b = ab \quad \text{and} \quad a \otimes b = a^{\log_q b}.$$

Show that P is a field with respect to the operations $\oplus$ and $\otimes$.

2. Further Examples of Groups

Suppose that we have a square as shown below,

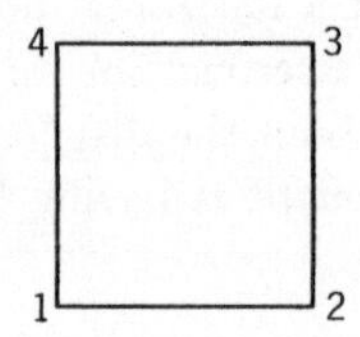

where we have numbered the vertices for convenience. We ask what we may "do" to the square that does not change its appearance. For example, if we perform a rotation, R, of the square through $90°$ in a counterclockwise direction we obtain

which, except for the labels on the vertices, is the original square. We may indicate R symbolically by giving its effect on the vertices thus:

$$R = \begin{pmatrix} 1 & 2 & 3 & 4 \\ 2 & 3 & 4 & 1 \end{pmatrix}$$

That is, the vertex 1 is where the vertex 2 was in the original square; the vertex 2 is where the vertex 3 was, and so on. Similarly, we have

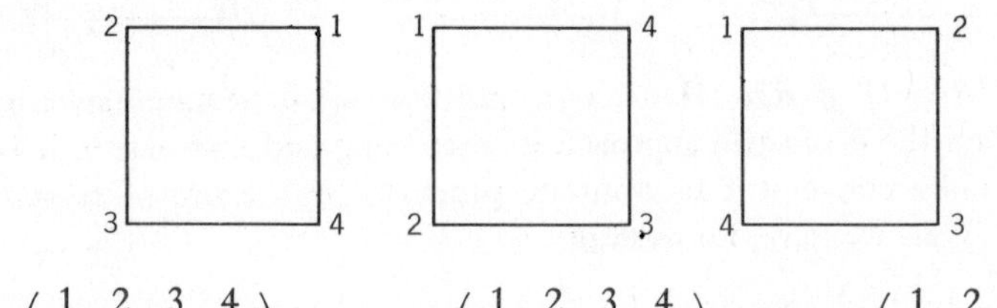

$$R' = \begin{pmatrix} 1 & 2 & 3 & 4 \\ 3 & 4 & 1 & 2 \end{pmatrix} \qquad R'' = \begin{pmatrix} 1 & 2 & 3 & 4 \\ 4 & 1 & 2 & 3 \end{pmatrix} \qquad H = \begin{pmatrix} 1 & 2 & 3 & 4 \\ 4 & 3 & 2 & 1 \end{pmatrix}$$

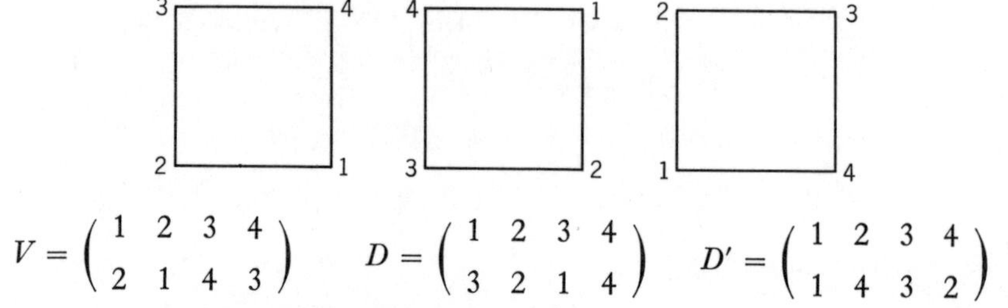

$$V = \begin{pmatrix} 1 & 2 & 3 & 4 \\ 2 & 1 & 4 & 3 \end{pmatrix} \qquad D = \begin{pmatrix} 1 & 2 & 3 & 4 \\ 3 & 2 & 1 & 4 \end{pmatrix} \qquad D' = \begin{pmatrix} 1 & 2 & 3 & 4 \\ 1 & 4 & 3 & 2 \end{pmatrix}$$

Then R' corresponds to a rotation through 180°; R'' to a rotation through 270°; H to a reflection about a horizontal line through the middle of the square; V to a reflection about a vertical line through the middle of the square; and D and D' to reflections about the diagonals. Finally, we may consider the original position of the square as having been obtained by the identity transformation,

$$I = \begin{pmatrix} 1 & 2 & 3 & 4 \\ 1 & 2 & 3 & 4 \end{pmatrix}$$

Let us now consider the result of applying first R and then H to the square. We have

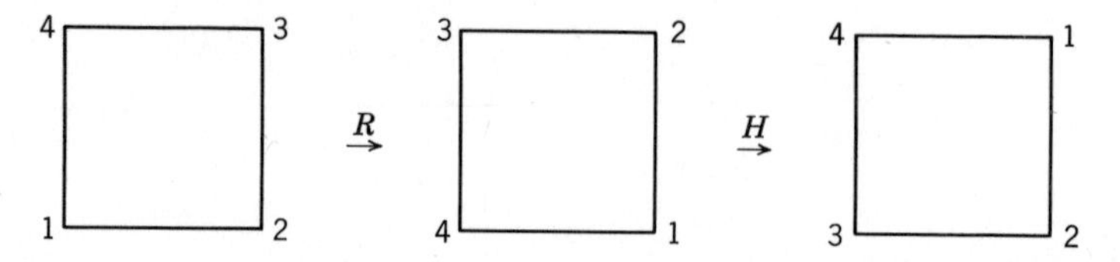

so that the effect of R followed by H is the same as the effect of D. We write $RH\ (= R \circ H) = D$. On the other hand, we have

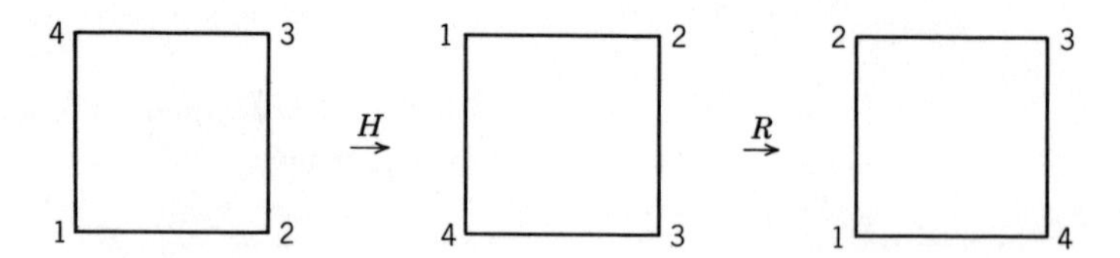

so that $HR = D' \neq RH$. Hence our operation is a noncommutative one.

Although the geometric approach is interesting and instructive, it is frequently more convenient to compute products by the use of *permutation symbols*. Thus we have, for example,

$$RH = \begin{pmatrix} 1 & 2 & 3 & 4 \\ 2 & 3 & 4 & 1 \end{pmatrix}\begin{pmatrix} 1 & 2 & 3 & 4 \\ 4 & 3 & 2 & 1 \end{pmatrix} = \begin{pmatrix} 1 & 2 & 3 & 4 \\ 3 & 2 & 1 & 4 \end{pmatrix} = D$$

since we have $1 \xrightarrow{R} 2 \xrightarrow{H} 3$ and therefore $1 \xrightarrow{RH} 3$; $2 \xrightarrow{R} 3 \xrightarrow{H} 2$ and therefore $2 \xrightarrow{RH} 2$; $3 \xrightarrow{R} 4 \xrightarrow{H} 1$ and therefore $3 \xrightarrow{RH} 1$; $4 \xrightarrow{R} 1 \xrightarrow{H} 4$ and therefore $4 \xrightarrow{RH} 4$.

Similarly we have

$$HR = \begin{pmatrix} 1 & 2 & 3 & 4 \\ 4 & 3 & 2 & 1 \end{pmatrix}\begin{pmatrix} 1 & 2 & 3 & 4 \\ 2 & 3 & 4 & 1 \end{pmatrix} = \begin{pmatrix} 1 & 2 & 3 & 4 \\ 1 & 4 & 3 & 2 \end{pmatrix} = D'$$

Clearly $IR = R = RI$, $HI = IH = H$ and so on, and $I = II = RR'' = R''R = R'R' = HH = VV = DD = D'D'$.

The results obtained so far can be used to begin a "multiplication table" or, as it is sometimes called, a *Cayley array*.†

	I	R	R'	R''	H	V	D	D'
I	I	R	R'	R''	H	V	D	D'
R	R			I	D			
R'	R'		I					
R''	R''	I						
H	H	D'			I			
V	V					I		
D	D						I	
D'	D'							I

If the student will now compute the remaining products he will see that we certainly have closure, a right-hand (and also left-hand) identity, and right-hand (and also left-hand) inverses for every element. Thus only the verification of the associative property is needed to proclaim the fact that we have a (noncommutative) group, called the *group of symmetries of the square*. Unfortunately, however, the fact that our operation is indeed associative cannot be readily seen from the table and to check the associative property case by case is much too lengthy a procedure. Later we will show that the multiplication of permutations is always associative but for the moment we will simply assume this fact.

† In honor of A. Cayley (1821–1895) who was one of the first mathematicians to conduct extensive research in the theory of groups.

Exercises 7.2

1. Complete the Cayley array for the group of symmetries of the square.
2. Construct the group of symmetries of an equilateral triangle.
3. Construct the group of symmetries of a rectangle that is not a square.
*4. Construct the group of symmetries of a regular tetrahedron.

3. Some Simple Properties of Groups

It will be useful to make first the following definition.

Definition 4. *If a right identity e is also a left identity, that is if $ea = a$ for all a in G, we say that e is an* ***identity*** *of G. If a' is a right inverse of a and if, also, a' is a left inverse of a, i.e., if $a'a = e$, then we will say that a' is an* ***inverse*** *of a.*

We now list several simple but important properties of groups in the following theorem.

Theorem 2. Let G be a group with a right identity e and suppose a, b, and $c \in G$ with a' a right hand inverse of a. Then
(1) $ba = ca$ implies $b = c$ (right-hand cancellation law);
(2) $ea = a$ for all $a \in G$ (i.e., a right identity is also a left identity);
(3) $a'a = e$ (that is a right inverse is also a left inverse);
(4) $ab = ac$ implies $b = c$ (left-hand cancellation law);
(5) the equations $ax = b$ and $ya = b$ have unique solutions x and $y \in G$
(6) a group has one and only one right identity which is also the unique left identity and the unique identity;
(7) any element of G has one and only one right inverse which is also its unique left inverse and its unique inverse;
(8) a is the inverse of a';
(9) the inverse of ab is $b'a'$.

Proof.
(1) If $ba = ca$, then $(ba)a' = (ca)a'$. Hence $b(aa') = c(aa')$, $be = ce$, and $b = c$.
(2) We have $e(aa') = ee = e = aa'$. Hence $(ea)a' = aa'$ and, by (1), $ea = a$.
(3) Since $aa' = e$, it follows that $a'(aa') = a'e = a'$. But $a' = ea'$ by (2). Hence $a'(aa') = ea'$. Since $a'(aa') = (a'a)a'$ we have $(a'a)a' = ea'$. Again applying (1) we have $a'a = e$.
(4) By (3) we know that there exists $a' \in G$ such that $a'a = e$. Then $ab = ac$ implies $a'(ab) = a'(ac)$. Thus $(a'a)b = (a'a)c$ so that $eb = ec$, and $b = c$.
(5) If $x = a'b$ and $y = ba'$ we have $ax = a(a'b) = (aa')b = eb = b$ and $ya = (ba')a = b(a'a) = be = b$. Hence there are solutions. If x' and y' are also solutions of the respective equations we have $ax = ax'$ and $ya = y'a$. But then $x = x'$ and $y = y'$ by (4) and (1) respectively.

(6) A right identity may be considered as a solution of the equation $ax = a$ and is therefore unique by (5). By (2) and Definition 4 it follows that this unique right identity is also the unique left identity and the unique identity.

(7) A right inverse of an element a may be considered as a solution of the equation $ax = e$ and is therefore unique by (5). By (3) and Definition 4 if follows that this unique right inverse of a is also the unique left inverse of a and the unique inverse of a.

(8) By (3), the inverse a' of a has the property that $a'a = aa' = e$ and hence a is the inverse of a'.

(9) $(ab)(b'a') = a[b(b'a')] = a[(bb')a'] = a(ea') = aa' = e$.

We will denote the unique inverse of a by a^{-1} from now on and define a^{-n} for $n \in N$ as $(a^{-1})^n$.

Exercises 7.3

1. If a, b, and c are elements of a group, prove that the equation $bxaxba = bxc$ has a unique solution.
2. Prove that, if x is an element of a group and $xx = x$, then $x = e$, the identity element of the group.
3. Prove that $(ab)(ab) = (aa)(bb)$ for all elements a and b of a group if and only if the group is Abelian.
4. Prove that if G is a group with identity e such that $a \in G$ implies $a^2 = e$, then G is an Abelian group.
5. In a group with an even number of elements, show that there is an element besides the identity that is its own inverse.
6. Let S be a set of elements, $a, b, c, \ldots$ and $\circ$ a binary operation on S satisfying (3) of Definition 1, and in addition, having the property that all equations $x \circ a = b$ and $a \circ y = b$ have solutions x and y in S. Prove that S is a group.

*7. If x is an element of a group G and there exists $y \in G$ such that $y^2 = x$ we will say that x has a square root. Prove (1) if G is a finite group and each $a \in G$ has a square root, then the root is unique and (2) each element of a finite group has a square root if and only if the order of the group is odd.

*8. Let $a_1, a_2, \ldots, a_n$ be n, not necessarily distinct, elements of a group G of order n. Show that there exist integers p and q, $1 \leq p \leq q \leq n$ such that $a_p a_{p+1} \ldots a_q = e$, the identity of G.

*9. Prove that if G is a group with identity e and having more than two elements, then there exist $a, b \in G$ with $a \neq b$, $a \neq e$, $b \neq e$ such that $ab = ba$.

4. Permutations

We have already given an example of a group of permutations in considering the symmetries of a square. The generalization is simple. By

$$p = \begin{pmatrix} 1 & 2 & \ldots & n \\ j_1 & j_2 & \ldots & j_n \end{pmatrix}$$

we mean the permutation which sends 1 into j_1, 2 into j_2, ... , n into j_n where $j_1, j_2, \ldots, j_n$ is any arrangement of the integers 1, 2, ... , n. Note that the order of the columns in the symbol is immaterial. Thus, for example,

$$\begin{pmatrix} 1 & 2 & 3 \\ 3 & 1 & 2 \end{pmatrix} = \begin{pmatrix} 2 & 1 & 3 \\ 1 & 3 & 2 \end{pmatrix}$$

Likewise, our definition of multiplication is a generalization of the multiplication used in our example: If

$$q = \begin{pmatrix} j_1 & j_2 & \ldots & j_n \\ k_1 & k_2 & \ldots & k_n \end{pmatrix}$$

we have

$$pq = \begin{pmatrix} 1 & 2 & \ldots & n \\ k_1 & k_2 & \ldots & k_n \end{pmatrix}$$

We have observed that multiplication of permutations is not commutative. We now prove that, however, it is associative. For if

$$r = \begin{pmatrix} k_1 & k_2 & \ldots & k_n \\ m_1 & m_2 & \ldots & m_n \end{pmatrix}$$

we have

$$(pq)r = \begin{pmatrix} 1 & 2 & \ldots & n \\ k_1 & k_2 & \ldots & k_n \end{pmatrix}\begin{pmatrix} k_1 & k_2 & \ldots & k_n \\ m_1 & m_2 & \ldots & m_n \end{pmatrix} = \begin{pmatrix} 1 & 2 & \ldots & n \\ m_1 & m_2 & \ldots & m_n \end{pmatrix}$$

and also

$$p(qr) = \begin{pmatrix} 1 & 2 & \ldots & n \\ j_1 & j_2 & \ldots & j_n \end{pmatrix}\begin{pmatrix} j_1 & j_2 & \ldots & j_n \\ m_1 & m_2 & \ldots & m_n \end{pmatrix} = \begin{pmatrix} 1 & 2 & \ldots & n \\ m_1 & m_2 & \ldots & m_n \end{pmatrix}$$

Furthermore, we have the identity permutation

$$\begin{pmatrix} 1 & 2 & \ldots & n \\ 1 & 2 & \ldots & n \end{pmatrix}$$

for any n symbols and, relative to this identity permutation, we have

$$p^{-1} = \begin{pmatrix} j_1 & j_2 & \ldots & j_n \\ 1 & 2 & \ldots & n \end{pmatrix}$$

Now let us consider the set, S_n, of all $n!$ permutations on n symbols. Our remarks above give us the following theorem.

Theorem 3. S_n is a group with respect to permutation multiplication.

Definition 5. *S_n is called the* ***symmetric group*** *on n symbols.*

It is often convenient to use the *cyclic notation* for permutations in which, for example, the permutation

$$\begin{pmatrix} 1 & 2 & 3 & 4 \\ 3 & 1 & 4 & 2 \end{pmatrix}$$

is written as (1 3 4 2). We read the new symbol, called a *cycle*, as follows: 1 is replaced by 3, 3 is replaced by 4, 4 is replaced by 2, and 2 is replaced by 1. Note that (1 3 4 2) = (3 4 2 1) = (4 2 1 3) = (2 1 3 4). Similarly, the permutation

$$\begin{pmatrix} 1 & 2 & 3 & 4 & 5 \\ 2 & 3 & 1 & 5 & 4 \end{pmatrix}$$

is written as (1 2 3)(4 5) and the permutation

$$\begin{pmatrix} 1 & 2 & 3 \\ 3 & 2 & 1 \end{pmatrix}$$

as (1 3)(2) or simply as (1 3) where it is understood that when a symbol is omitted it remains fixed in the permutation.

Definition 6. *Two cycles are said to be* ***disjoint*** *if they have no symbols in common.*

Thus (1 3 4) and (2 5) are disjoint but (1 3 4) and (2 3 5) are not disjoint.

From the procedure used in writing permutations in terms of cycles, the following result is now evident.

Theorem 4. Every permutation can be written as the product of disjoint cycles.

Example. Express (1 3 2 5)(1 4 3)(2 5 1) as the product of disjoint cycles.

Under the first cycle, 1 goes to 3 but then, under the second cycle 3 goes to 1. Finally, under the third cycle 1 goes to 2. Thus we have (1 2 ...). Now under the first cycle 2 goes to 5 and then, under the third cycle, 5 goes to 1. Thus we have (1 2).... Now we look at what happens to 3. First 3 goes to 2 (first cycle). Then 2 goes to 5 (third cycle) so we have (1 2)(3 5 .... Now 5 goes to 1 (first cycle) and 1 goes to 4 (second cycle). Thus we have (1 2)(3 5 4 .... Now 4 goes to 3 (second cycle) so we have (1 2)(3 5 4). That is, (1 3 2 5)(1 4 3)(2 5 1) = (1 2)(3 5 4).

Definition 7. *A permutation that can be written as a cycle on just two symbols, $(i\,j)$, is called a* ***transposition.***

Theorem 5. Any permutation can be written as a product of transpositions.

Proof. We first observe that a cycle of n symbols can be written as a product of transpositions since

$$(1\ 2\ 3\ 4\ \dots\ n) = (1\ 2)(1\ 3)(1\ 4)\ \dots\ (1\ n)$$

Our result then follows from Theorem 4.

We note that the expression of a permutation as a product of transpositions is by no means unique. For example, $(1\ 2\ 3) = (1\ 2)(1\ 3) = (1\ 3)(1\ 2)(1\ 3)(1\ 2)$. In both cases, however, we observe that the number of transpositions is even.

Theorem 6. If p is a product of s transpositions and also a product of r transpositions, then $r \equiv s \pmod 2$; i.e., r and s are either both odd or both even.†

Proof. Our proof uses the *alternating polynomial*, A, in the n symbols $x_1, x_2, \dots, x_n$. It is defined as the product of the $n(n-1)/2$ factors $x_i - x_j$ where $i < j$. Thus if $n = 4$

$$A = (x_1 - x_2)(x_1 - x_3)(x_1 - x_4)(x_2 - x_3)(x_2 - x_4)(x_3 - x_4)$$

In general,

$$\begin{aligned} A &= \prod_{i<j}^{n} (x_i - x_j) \\ &= (x_1 - x_2)(x_1 - x_3)(x_1 - x_4)\dots(x_1 - x_n) \\ &\qquad (x_2 - x_3)(x_2 - x_4)\dots(x_2 - x_n) \\ &\qquad\qquad (x_3 - x_4)\dots(x_3 - x_n) \\ &\qquad\qquad \dots\dots\dots\dots\dots\dots \\ &\qquad\qquad\qquad (x_{n-1} - x_n) \end{aligned}$$

Consider now any permutation, p, of $1, 2, \dots, n$. By Ap we mean the polynomial obtained by permuting the subscripts $1, 2, \dots, n$ of the x_i as prescribed by p. For example, if $n = 4$ and $p = (1\ 3\ 4\ 2)$,

$$Ap = (x_3 - x_1)(x_3 - x_4)(x_3 - x_2)(x_1 - x_4)(x_1 - x_2)(x_4 - x_2)$$

In particular, if $t = (i\ j)$ where $i < j$, $At = -A$. For, the factors $x_r - x_s$ of A with $r \neq i$ or j and $s \neq i$ or j remain unchanged and $x_i - x_j$ becomes $x_j - x_i = -(x_i - x_j)$. The remaining factors of A are of the form $x_r - x_s$ where either r or s, but not both, is equal to i or j. In particular, they are

(1) $x_1 - x_i, x_2 - x_i, \dots, x_{i-1} - x_i$

(2) $x_i - x_{i+1}, x_i - x_{i+2}, \dots, x_i - x_{j-1}$

† In such a case we say that r and s have the same *parity*.

(3) $x_i - x_{j+1}, x_i - x_{j+2}, \ldots, x_i - x_n$

$x_1 - x_j, x_2 - x_j, \ldots, x_{i-1} - x_j$

(5) $x_{i+1} - x_j, x_{i+2} - x_j, \ldots, x_{j-1} - x_j$

(6) $x_j - x_{j+1}, x_j - x_{j+2}, \ldots, x_j - x_n$

But then the $i - 1$ factors in (1) may be paired with the $i - 1$ factors in (4) as follows:

(7) $(x_1 - x_i)(x_1 - x_j), \quad (x_2 - x_i)(x_2 - x_j), \ldots, (x_{i-1} - x_i)(x_{i-1} - x_j)$

the $j - i - 1$ factors in (2) may be paired with the $j - i - 1$ factors in (5) as follows:

(8) $(x_i - x_{i+1})(x_{i+1} - x_j), (x_i - x_{i+2})(x_{i+2} - x_j), \ldots, (x_i - x_{j-1})(x_{j-1} - x_j)$

and the $n - j$ factors in (3) may be paired with the $n - j$ factors in (6) as follows:

(9) $(x_i - x_{j+1})(x_j - x_{j+1}), (x_i - x_{j+2})(x_j - x_{j+2}), \ldots, (x_i - x_n)(x_j - x_n)$.

But in (7), $(x_k - x_i)(x_k - x_j)t = (x_k - x_j)(x_k - x_i) = (x_k - x_i)(x_k - x_j)$ for $k = 1, 2, \ldots, i - 1$; in (8), $(x_i - x_k)(x_k - x_j)t = (x_j - x_k)(x_k - x_i) = (x_i - x_k)(x_k - x_j)$ for $k = i + 1, \ldots, j - 1$; and in (9), $(x_i - x_k)(x_j - x_k)t = (x_j - x_k)(x_i - x_k) = (x_i - x_k)(x_j - x_k)$ for $k = j + 1, \ldots, n$.

(The student should check that there are just three special cases in which the tabulation of factors given in (1) to (6) breaks down. These cases are: (a) When $i = 1$, in which case factors of the form (1) and (4) do not appear; (b) when $j = n$, in which case factors of the form (3) and (6) do not appear; and (c) when $j = i + 1$, in which case factors of the form (2) and (5) do not appear. Because of the pairing of factors in (1) and (4), (3) and (6), and (2) and (5), however, these exceptional cases do not invalidate our result.)

Thus the total effect of p on A is simply to change $x_i - x_j$ to $x_j - x_i$ and hence $At = -A$.

Now suppose p is a product of r transpositions. Then $Ap = (-1)^r A$. Likewise, if p is the product of s transpositions, $Ap = (-1)^s A$. Hence $(-1)^r = (-1)^s$, $r \equiv s \pmod 2$.

Definition 8. *If a permutation is the product of an even number of transpositions it is called an* ***even*** *permutation; otherwise it is called an* ***odd*** *permutation.*

Theorem 7. Of the $n!$ permutations on n symbols, $n!/2$ are even permutations and $n!/2$ are odd permutations.

Proof. Let $e_1, e_2, \ldots, e_r$ be the even permutations and $o_1, o_2, \ldots, o_s$ be the odd permutations where, of course, $r + s = n!$ Now let t be any transposition. Since t is obviously an odd permutation it follows that $te_1, te_2, \ldots, te_r$ are odd permutations and that $to_1, to_2, \ldots, to_s$ are even permutations. Now clearly no $te_i = to_j$ for any $i = 1, \ldots, r, j = 1, \ldots, s$ since an odd permutation is never also an even permutation. But, furthermore, if $te_i = te_j$, then $e_i = e_j$ by the cancellation law. Similarly $to_i \neq to_j$ if $i \neq j$. Thus all of the r even permutations must appear in the list $to_1, \ldots, to_s$ and all of the s odd permutations in the list $te_1, \ldots, te_r$. Hence $r = s = n!/2$.

Exercises 7.4

1. Write the 3! permutations on three symbols in the two-rowed notation; as a product of disjoint cycles; as a product of transpositions. Which are even permutations and which are odd permutations?

2. Express each of the following products of permutations on six symbols as a product of disjoint cycles. Also find the inverse of each product and express it as a product of disjoint cycles.
   (a) $(2\ 3\ 4)(2\ 4)$ (b) $(1\ 3\ 2\ 4)(5\ 2)$
   (c) $(1\ 3)(2\ 4)(2\ 3)$ (d) $(1\ 4\ 3\ 1)(2\ 5\ 2)(1\ 3)$
   (e) $(1\ 4\ 3\ 2)(2\ 4\ 1)(1\ 3\ 5)$ (f) $(2\ 3)(1\ 4)(2\ 5)$

3. Prove that a cycle containing an odd number of symbols is an even permutation whereas a cycle containing an even number of symbols is an odd permutation.

4. Prove that the inverse of an even permutation is an even permutation and the inverse of an odd permutation is an odd permutation.

5. Prove that every even permutation is either a cycle of length three or can be expressed as a product of cycles of length three.

6. Prove that the $n!/2$ even permutations on n symbols form a group with respect to permutation multiplication. (This group is known as the *alternating group* on n symbols and is designated by A_n.)

5. Mappings

Roughly speaking, a mapping (or function) from one set, S, into another, T, is a rule that associates with each element $s \in S$ a unique element $t \in T$. For example, if S is the set of real numbers and T the set of non-negative real numbers we may consider the mapping that associates $x^2 \in T$ with $x \in S$; $x \in S \to x^2 \in T$.

For a more precise definition, recall that (Section 1.4) we have defined the Cartesian product of two sets, $S \times T$, by $S \times T = \{(s, t) | s \in S, t \in T\}$.

Definition 9. *If S and T are nonempty sets, then a* ***mapping*** *from S into T is a subset, M, of $S \times T$ such that for every $s \in S$ there is a unique $t \in T$ such that the ordered pair (s, t) is in M.*

In terms of this definition, then, the mapping defined above *is* the subset M of $S \times T$. Nevertheless, we find it more useful and convenient (as do most mathematicians) to continue to think (somewhat loosely) of a mapping as a rule which associates with any element in S some element in T, the rule being to associate $s \in S$ with t in T if and only if $(s, t) \in M$. We call t the *image* of s under the mapping.

Now we need some notation. Let σ be a mapping from S into T; we write

$$\sigma : S \to T \quad \text{or} \quad S \overset{\sigma}{\to} T$$

If $t \in T$ is the image of $s \in S$ under σ we may also write

$$\sigma : s \to t \quad \text{or} \quad s \overset{\sigma}{\to} t$$

Still other notations are, for $s \in S$ and $t \in T$ the image of s,

$$t = s\sigma \quad \text{or} \quad t = \sigma(s)$$

The notation $t = s\sigma$ is the one preferred by most algebraists; the notation $t = \sigma(s)$ by most other mathematicians. Since the latter notation is more familiar in terms of the customary functional notation we shall utilize it sometimes but will also use the notation $t = s\sigma$ when convenient.

Example 1. Consider any set $S \neq \varnothing$ and define $\iota : S \to S$ by $\iota(s) = s\iota = s$. The mapping ι is called the *identity* mapping of S into S.

Example 2. Let $S \neq \varnothing$ and $T \neq \varnothing$ be sets and t_o a fixed element of T. Then a mapping σ of S into T is defined by $\sigma : s \in S \to t_o \in T$.

Example 3. Let $S = N$ and $T = I$. Then a mapping τ of S into T is defined by $\tau : k \in S \to +k \in T$ (see Section 4.5).

Example 4. Let $S = I$ and $T = R$. Then a mapping σ of S into T is defined by $\sigma : a \in S \to a/1 \in T$ (see Section 6.4).

Example 5. Let $S = I \times I$, $T = I$, and define $\sigma : S \to T$ by $(m, n)\sigma = m + n$ for $m, n \in I$.

Example 6. Let $S = \{x_1, x_2, x_3\}$ and define $\tau : S \to S$ by $\tau(x_1) = x_2$, $\tau(x_2) = x_3$, $\tau(x_3) = x_1$.

Example 7. Let $S = I$ and $T = \{C_o, C_1\} = J_2$. Define $\sigma : S \to T$ by $\sigma(2n) = C_o$ and $\sigma(2n + 1) = C_1$ for $n \in I$.

There are two special kinds of mappings that are of great importance.

Definition 10. *A mapping σ of S into T is said to be* **onto** *T if for every $t \in T$, then exists an $s \in S$ such that $t = \sigma(s)$.*

The mapping of Example 1 is clearly onto; the mapping of Example 2 is onto if and only if $T = \{t_o\}$; the mapping of Example 3 is not onto since no element of S maps, for example, into -1. It is left as an exercise for the student to show that the mappings of Examples 5, 6, and 7 are onto but that the mapping of Example 4 is not onto.

Definition 11. *A mapping, σ, of S into T is said to be a* **one-to-one** *(1–1)* **mapping** *if whenever $s_1 \neq s_2$ for $s_1, s_2 \in S$, then $s_1\sigma \neq s_2\sigma$ or, equivalently, if $s_1\sigma = s_2\sigma$* **implies** *$s_1 = s_2$.*

The mapping of Example 1 is clearly 1–1 but, unless S has only element, the mapping of Example 2 is not 1–1. The mapping of Example 3 is 1–1 since if $s_1, s_2 \in N$, and $s_1 \neq s_2$ we have $s_1\sigma = +s_1 \neq s_2\sigma = +s_2$; the mapping of Example 5 is not 1–1 since, for example, $(1, 3) \neq (2, 2)$ but $(1, 3)\sigma = 1 + 3 = 4 = (2, 2)\sigma = 2 + 2$. It is left as an exercise to the student to show that the mappings of Examples 4 and 6 are 1–1 but that the mapping of Example 7 is not 1–1.

Exercises 7.5

1. If $S = \{a, b, c, d\}$ and $T = \{x, y\}$, form the Cartesian product, $S \times T$, of S and T.
2. Show that the mappings of Examples 5, 6, and 7 are onto but that the mapping of Example 4 is not onto.
3. Show that the mappings of Examples 4 and 6 are 1–1 but that the mapping of Example 7 is not 1–1.
4. For the following mappings, $\sigma : S \to T$, determine (1) if σ is onto and (2) whether σ is 1–1.
   (a) S = the set of real numbers, T = set of non-negative real numbers, $\sigma(s) = s^2$ for $s \in S$.
   (b) $S = T$ = set of real numbers, $\sigma(s) = s^3$ for $s \in S$.
   (c) $S = T$ = set of real numbers, $\sigma(s) = s^2$ for $s \in S$.
   (d) $S = T$ = set of real numbers, $\sigma(s) = s^5$ for $s \in S$.
   (e) $S = T = I$, $\sigma(s) = 2s$ for $s \in S$.
   (f) $S = \{x_1, x_2, x_3\}$, $T = \{y_1, y_2\}$, $\sigma(x_1) = y_1$, $\sigma(x_2) = y_1$, $\sigma(x_3) = y_2$.
   (g) $S = \{x_1, x_2, x_3\}$, $T = \{y_1, y_2, y_3\}$, $\sigma(x_1) = y_1$, $\sigma(x_2) = y_1$, $\sigma(x_3) = y_3$.
   (h) $S = I \times I$, $T = I$, $\sigma(m, n) = mn$ for $m, n \in I$.
   (i) $S = I$, $T = \{C_0, C_1, C_2\}$, $\sigma(3n) = C_0$, $\sigma(3n + 1) = C_1$, $\sigma(3n + 2) = C_2$ for $n \in I$.
5. If S and T are two nonempty sets prove that there exists a 1–1 map of $S \times T$ onto $T \times S$.

6. Isomorphisms and Automorphisms of Groups

We have already touched upon the concept of an isomorphism in Section 2.6. We now relate this concept to groups using the idea of a mapping as developed in the previous section.

Definition 12. *A group G with operation $\circ$ will be said to be* ***isomorphic*** *to a group G' with operation $*$ if there exists a 1–1 map σ of G onto G' such that $\sigma(a \circ b) = \sigma(a) * \sigma(b)$. We write $G \cong G'$.*

Definition 12 is simply a more formal and precise statement, as applied to groups, of our previous idea of an isomorphism as a 1–1 correspondence that is preserved under the operations involved. Consider, for example, our previous discussion of the isomorphism of I with the set $I' = \{a/1 \mid a \in I\} \subset R$ (Section 6.4). Our mapping is $\sigma : a \in I \rightarrow \sigma(a) = a/1 \in I'$. It is an onto mapping since any $a/1 \in I'$ is obtained from $a \in I$ because $\sigma(a) = a/1$; it is 1–1 since if $a \neq b$ but $\sigma(a) = \sigma(b)$ we would have $a/1 = b/1$, a contradiction. The correspondence is preserved under the operation of addition since $\sigma(a + b) = (a + b)/1 = a/1 + b/1 = \sigma(a) + \sigma(b)$ and under multiplication since $\sigma(ab) = (ab)/1 = (a/1)(b/1) = \sigma(a)\sigma(b)$.

Example 1. Let G be the group whose elements are 1 and -1 under multiplication and G' the additive group modulo 2 with elements C_0 and C_1. Then G is isomorphic to G' since the map σ defined by $\sigma(1) = C_0$ and $\sigma(-1) = C_1$ is obviously 1–1 and onto. Furthermore

$$\sigma(1 \cdot 1) = \sigma(1) = C_0 = C_0 + C_0 = \sigma(1) + \sigma(1)$$

$$\sigma(1 \cdot -1) = \sigma(-1) = C_1 = C_0 + C_1 = \sigma(1) + \sigma(-1)$$

$$\sigma(-1 \cdot 1) = \sigma(-1) = C_1 = C_1 + C_0 = \sigma(-1) + \sigma(1)$$

$$\sigma(-1 \cdot -1) = \sigma(1) = C_0 = C_1 + C_1 = \sigma(-1) + \sigma(-1)$$

Example 2. Let G be the group of all positive real numbers under multiplication and let G' be the set of all real numbers under addition. Then $G \cong G'$ under the map σ defined by $\sigma(a) = \log a$ for $a \in G$. (These logarithms can be to any fixed base. We will assume natural—base e—logarithms.) The fact that the correspondence is 1–1 and onto is a consequence of the fact that every positive real number has a unique logarithm and that every real number has a unique antilogarithm. Thus if $a \in G \neq b \in G$ then $\log a \in G' \neq \log b \in G'$ and for $x \in G'$ we have $x = \log e^x = \sigma(e^x)$ where $e^x \in G$.

Theorem 8. The relationship of isomorphism between groups is an equivalence relation.

Proof. (1) $G \cong G$ by virtue of the identity mapping $\iota : a \to a$.

(2) If $G \cong G'$, then $G' \cong G$. We let σ be the 1–1 mapping of G onto G'. Now we define a mapping τ of G' into G by $\tau(a') = a$ ($a' \in G'$ and $a \in G$) if and only if $\sigma(a) = a'$. (For example, if σ is the mapping of $\{1, -1\}$ onto $\{C_0, C_1\}$ defined by $1 \to C_0 = \sigma(1)$ and $-1 \to C_1 = \sigma(-1)$, then τ is the mapping of $\{C_0, C_1\}$ onto $\{1, -1\}$ defined by $C_0 \to 1 = \tau(C_0)$ and $C_1 \to -1 = \tau(C_1)$. Thus in the notation previously used we have $1 \leftrightarrow C_0$ and $-1 \leftrightarrow C_1$.)

Then τ is 1–1 since if $a' \in G' \neq b' \in G'$ but $\tau(a') = a = \tau(b') = b$ we have $a = b$ so that $\sigma(a) = a' = \sigma(b) = b'$, a contradiction. Furthermore, given any $a \in G$ we consider $\sigma(a) = a' \in G'$ so that $\tau(a') = a$ and hence conclude that τ is a mapping of G' onto G. Finally, if o is our operation in G and $*$ our operation in G', we have $\tau(a' * b') = \tau[\sigma(a) * \sigma(b)] = \tau[\sigma(a \circ b)] = \tau[(a \circ b)'] = a \circ b$ where we use the fact that G is isomorphic to G' under σ to write $\sigma(a) * \sigma(b) = \sigma(a \circ b)$.

(3) If $G \cong G'$ and $G' \cong G''$, then $G \cong G''$. We leave the proof of this as an exercise for the student.

Since $\cong$ for groups has been shown to be an equivalence relation we know that it partitions the set of all groups into classes of equivalent groups (Section 4.2). Roughly speaking, each member of a class of equivalent groups is the same except for notation. Furthermore, because of Theorem 8 we may speak of two groups G and G' being isomorphic rather than saying G is isomorphic to G' or G' is isomorphic to G. Thus the two groups $G = \{1, -1\}$ under multiplication and $G' = \{C_0, C_1\}$ under addition are isomorphic and the Cayley arrays below show clearly the idea that, essentially, they differ only in notation.

$\times$	1	-1
1	1	-1
-1	-1	1

$+$	C_0	C_1
C_0	C_0	C_1
C_1	C_1	C_0

When we are considering two groups G and G' abstractly we frequently do not bother to distinguish symbolically between the operations on the groups. Thus, instead of writing $a \circ b \in G$ and $a' * b' \in G'$, for example, we write, in both cases, ab and $a'b'$. We shall do this from now on. Also, we shall sometimes revert to our previous notation, $a \in G \leftrightarrow a' \in G'$, when no confusion can result from its use.

It is obvious, of course, that if two finite groups have different orders they cannot be isomorphic. On the other hand, our next example shows that two groups of the same order need not be isomorphic.

Example 3. Let $G = \{1, -1, i, -i\}$ (i the complex unit) under multiplication and let $G' = \{e, a, b, c\}$ with the group operation defined by the following Cayley array:

	e	a	b	c
e	e	a	b	c
a	a	e	c	b
b	b	c	e	a
c	c	b	a	e

Then G and G' are not isomorphic.

Before we investigate the reason for the non-isomorphism let us prove the following theorem:

Theorem 9. If $G \cong G'$ under the mapping $\sigma : a \in G \to a' \in G'$, e is the identity of G, and e' the identity of G', then $\sigma(e) = e'$ and if $\sigma(a) = a'$, then $\sigma(a^{-1}) = (a')^{-1}$.

Proof. (1) Suppose $\sigma(e) = a' \in G'$. We let b' be any element of G' and choose $b \in G$ so that $\sigma(b) = b'$. Then $\sigma(eb) = \sigma(b) = b'$ and also $\sigma(eb) = \sigma(e)\sigma(b) = a'b'$. Hence $a'b' = b'$ for all $b' \in G'$ and thus $a' = e'$ by uniqueness of the identity.

(2) Suppose $\sigma(a) = a'$ and $\sigma(a^{-1}) = b'$. Then $\sigma(aa^{-1}) = \sigma(e) = e'$ by (1). But also $\sigma(aa^{-1}) = \sigma(a)\sigma(a^{-1}) = a'b'$. Hence $a'b' = e'$ so that $b' = (a')^{-1}$ by uniqueness of inverses.

If we apply this theorem to Example 3 we may first conclude that if G is isomorphic to G' under the mapping σ, then $\sigma(1) = e$. Then we have the following three possibilities:

$$(1)\ \sigma(i) = a; \qquad (2)\ \sigma(i) = b; \qquad (3)\ \sigma(i) = c$$

Now if $\sigma(i) = a$, then $\sigma(i^{-1}) = a^{-1}$ by Theorem 9. However, $i^{-1} = -i$ and $a^{-1} = a$ so that $\sigma(-i) = a$. But this contradicts $\sigma(i) = a$ since our mapping must be 1–1. Since $b^{-1} = b$ and $c^{-1} = c$ we may similarly show that the two remaining possibilities are not permissible and hence $G \not\cong G'$.

It was fairly easy to show non-isomorphism in the example just discussed. In general, however, to decide whether or not two finite groups (of the same order) are isomorphic is not an easy task. It may require considerable ingenuity to avoid considering too long a list of possibilities.

Definition 13. *An isomorphism of a group G with itself is called an* ***automorphism*** *of G. That is, an automorphism of G is a 1–1 mapping of G onto G that is preserved under the group operation.*

Since an automorphism of a group is an isomorphism we must have $e \leftrightarrow e$ and, if $a \leftrightarrow b$, $a^{-1} \leftrightarrow b^{-1}$.

Example 4. Every group possesses at least one automorphism, called the *identity automorphism*, in which $a \in G \leftrightarrow a \in G$.

Example 5. In $G = \{1, -1, i, -i\}$ under multiplication, define σ by $\sigma(1) = 1$, $\sigma(-1) = -1$, $\sigma(i) = -i$, and $\sigma(-i) = i$. This mapping defines an automorphism of G since it is clearly 1–1 and onto and, for example, $\sigma[(-1)i] = \sigma(-i) = i = (-1)(-i) = \sigma(-1)\,\sigma(i)$; $\sigma[(-1)(-i)] = \sigma(i) = -i = (-1)i = \sigma(-1)\sigma(-i)$, etc.

Example 6. For G as in Example 5, the mapping defined by $\tau(1) = 1$, $\tau(-1) = i$, $\tau(i) = -1$, and $\tau(-i) = -i$ does not define an automorphism since $\tau[(-1)^{-1}] = \tau(-1) = i \neq [\tau(-1)]^{-1} = i^{-1} = -i$.

An important special class of automorphisms is described in the following theorem.

Theorem 10. Let G be a group and g any element of G. The mapping $\sigma_g : a \in G \to g\,a\,g^{-1}$ is an automorphism of G.

Proof. The mapping is 1–1 since if $\sigma_g(a) = g\,a\,g^{-1} = \sigma_g(b) = g\,b\,g^{-1}$ we have $a = b$. It is onto since if $b \in G$, $\sigma_g(g^{-1}\,b\,g) = g(g^{-1}\,b\,g)g^{-1} = b$. Finally, $\sigma_g(ab) = g(ab)g^{-1} = (g\,a\,g^{-1})(g\,b\,g^{-1}) = \sigma_g(a)\sigma_g(b)$.

> ***Definition 14.*** *An automorphism σ defined on a group G by $\sigma(a) = g\,a\,g^{-1}$ for a fixed $g \in G$ is called an* ***inner automorphism*** *of G. Any automorphism of a group that is not an inner automorphism is called an* ***outer automorphism****. The element $g\,a\,g^{-1}$ is called the* ***conjugate*** *of a under g.*

We now define what we mean by two equal isomorphisms and, more generally, two equal mappings.

> ***Definition 15.*** *Two mappings σ and τ of S into T are said to be* ***equal*** *if $s\sigma = s\tau$ for every $s \in S$.*

Notice that if g commutes with all elements of G, the inner automorphism σ_g is the identity automorphism. In particular, if G is Abelian, the only inner automorphism of G is the identity automorphism.

Example 7. The group of symmetries of a square (Section 2) has four distinct inner automorphisms. To see this we observe first that there are certainly at most eight inner automorphisms—one for each element of the group. They can be calculated by computing $RIR^{-1}, RRR^{-1}, \ldots, RD'R^{-1}$; $R'IR'^{-1}, R'RR'^{-1}, \ldots, R'D'R'^{-1}; \ldots; D'ID'^{-1}, D'RD'^{-1}, \ldots, D'D'D'^{-1}$. Our results may be summarized in the following table:

	I	R	R'	R''	H	V	D	D'
σ_I	I	R	R'	R''	H	V	D	D'
σ_R	I	R	R'	R''	V	H	D'	D
σ_H	I	R''	R'	R	H	V	D'	D
σ_D	I	R''	R'	R	V	H	D	D'

Thus, for example, the second line of the table means that

$$I \overset{\sigma_R}{\rightarrow} I,\ R \overset{\sigma_R}{\rightarrow} R,\ R' \overset{\sigma_R}{\rightarrow} R',\ R'' \overset{\sigma_R}{\rightarrow} R'',\ H \overset{\sigma_R}{\rightarrow} V,\ V \overset{\sigma_R}{\rightarrow} H, D \overset{\sigma_R}{\rightarrow} D', \text{ and } D' \overset{\sigma_R}{\rightarrow} D.$$

The student should check that $\sigma_I = \sigma_{R'}$, $\sigma_R = \sigma_{R''}$, $\sigma_H = \sigma_V$, and $\sigma_D = \sigma_{D'}$.

Exercises 7.6

1. Prove part (3) of Theorem 8.
2. Prove that $G = \{1, 5, 7, 11\}$ is a group under multiplication modulo 12 and that $G' = \{I, R', H, V\}$ (Section 2) is a group. Then (1) prove that $G \cong G'$; (2) G is not isomorphic to the group consisting of $1, -1, i, -i$ and group operation of multiplication.
3. Let $G = \{1, (-1 + i\sqrt{3})/2, (-1 - i\sqrt{3})/2\}$. Prove that G is a group under multiplication. Is it isomorphic to the group J_3 under addition?
4. Consider the group G of integers under addition and the group G' of even integers under addition. Is $G \cong G'$?
5. Find all of the automorphisms of the group of permutations $\{i = (1)(2)(3)(4), (12)(34), (13)(24), (14)(23)\}$. (This group is called the *fours group*.)

*6. Find all of the automorphisms of the symmetric group on three symbols. Are any of the automorphisms outer automorphisms?

*7. The elements of a field form a group under addition called the additive group of the field; the elements of a field with the exception of the additive identity form a group under multiplication called the multiplicative group of the field. Does there exist a field whose additive and multiplicative groups are isomorphic?

7. Automorphisms of a Field

Our discussion of automorphisms of fields will, to begin with, closely parallel our discussion of automorphisms of groups.

Definition 16. *An **automorphism** of a field F is an isomorphism of F with itself. More specifically, an automorphism of F is a 1–1 mapping of F onto F which is preserved under multiplication and addition.*

Example 1. Let F be any field and, for any $a \in F$, let $a \leftrightarrow a$. This trivial automorphism is called the *identity* automorphism.

Example 2. Let $F = \{a + b\sqrt{2} | a, b \in R\}$. Then $\sigma : a + b\sqrt{2} \to a + (-b)\sqrt{2} = a - b\sqrt{2}$ is an automorphism of F.

Proof. First of all, the mapping is 1–1 since if $\sigma(a + b\sqrt{2}) = a - b\sqrt{2} = \sigma(c + d\sqrt{2}) = c - d\sqrt{2}$, we have $a - b\sqrt{2} = c - d\sqrt{2}$ where a, b, c, $d \in R$. Hence if $b \neq d$, $\sqrt{2} = (a - c)/(b - d)$. But $\sqrt{2}$ is not a rational number (see Chapter 8) and hence $b = d$ from which it follows that $a = b$ and $a + b\sqrt{2} = c + d\sqrt{2}$. Thus σ is a 1–1 mapping. Furthermore, if $a + b\sqrt{2} \in F$, then $\sigma[a + (-b)\sqrt{2}] = a - (-b)\sqrt{2} = a + b\sqrt{2}$ so that σ is an onto mapping.

Now consider

$$\begin{aligned}\sigma[(a + b\sqrt{2}) + (c + d\sqrt{2})] &= \sigma[(a + c) + (b + d)\sqrt{2}] = \\ (a + c) - (b + d)\sqrt{2} &= (a - b\sqrt{2}) + (c - d\sqrt{2}) = \\ &= \sigma(a + b\sqrt{2}) + \sigma(c + d\sqrt{2})\end{aligned}$$

from which we conclude that our mapping is preserved under addition. Finally, we note that

$$\begin{aligned}\sigma[(a + b\sqrt{2})(c + d\sqrt{2})] &= \sigma[(ac + 2bd) + (bc + ad)\sqrt{2}] \\ (ac + 2bd) - (bc + ad)\sqrt{2} &= (a - b\sqrt{2})(c - d\sqrt{2}) \\ &= \sigma(a + b\sqrt{2})\sigma(c + d\sqrt{2})\end{aligned}$$

from which we conclude that our mapping is also preserved under multiplication.

Example 3. In R let $\sigma(a) = -a$. Clearly σ is a 1–1 and onto mapping but it is not an automorphism of R. We do, indeed, have $\sigma(a + b) = -(a + b) = (-a) + (-b) = \sigma(a) + \sigma(b)$ so that the mapping is preserved under addition but $\sigma(ab) = -(ab) \neq \sigma(a)\sigma(b) = (-a)(-b)$ so that the mapping is not preserved under multiplication.

The failure of the mapping in Example 3 to define an automorphism may be thought of as a consequence of the following theorem.

Theorem 11. In any automorphism, σ, of a field F, $\sigma(0) = 0$ and $\sigma(1) = 1$ where 0 is the additive identity of F and 1 is its multiplicative identity.

Proof. If $\sigma(0) = a$, then $\sigma(0 + 0) = \sigma(0) = a = \sigma(0) + \sigma(0) = a + a$. Hence $a + a = a$ and $a = 0$. Similarly, if $\sigma(1) = a$, $\sigma(1 \cdot 1) = \sigma(1) = a = \sigma(1)\sigma(1) = a \cdot a = a^2$. Hence $a^2 = a$ so that $a^2 - a = a(a - 1) = 0$ Thus $a - 1 = 0$ and $a = 1$ or $a = 0$. But we have shown that $\sigma(0) = 0$ and hence $\sigma(1) \neq 0$. Thus $\sigma(1) = 1$.

We want now to consider the product (or composition) of two mappings and, in particular, to apply this concept to automorphisms.

Definition 17. *If $\sigma : S \to T$ and $\tau : T \to U$ are mappings of S into T and T into U respectively, we define the* **product** *(or* **composition***), $\sigma \circ \tau$, of σ and τ by $\sigma \circ \tau : S \to U$ where $s(\sigma \circ \tau) = (s\sigma)\tau$ for all $s \in S$.*

Note that $\sigma \circ \tau$ means to apply σ first and then τ. If we were to use the notation $(\sigma \circ \tau)(s)$ we would define $(\sigma \circ \tau)(s) = \sigma[\tau(s)]$; i.e., τ would be applied first and then σ. It does not matter which convention is used but it certainly is important in reading the literature to know *which* convention is being employed. From now on we will always use the agreement made in Definition 17.

Example 4. Let $T = U = S = \{x_1, x_2, x_3\}$, define $\sigma : S \to T$ by $x_1\sigma = x_2$ $x_2\sigma = x_1$, and $x_3\sigma = x_3$, and define τ by $x_1\tau = x_1$, $x_2\tau = x_3$, and $x_3\tau = x_2$ Then $x_1(\sigma \circ \tau) = (x_1\sigma)\tau = x_2\tau = x_3$, $x_2(\sigma \circ \tau) = (x_2\sigma)\tau = x_1\tau = x_1$, and $x_3(\sigma \circ \tau) = (x_3\sigma)\tau = x_3\tau = x_2$. Note that $x_1(\tau \circ \sigma) = (x_1\tau)\sigma = x_1\sigma = x_2$ $\neq x_3 = x_1(\sigma \circ \tau)$ so that, in general, $\sigma \circ \tau \neq \tau \circ \sigma$.

Example 5. Let $S = R$, $T = I$, and $U = \{C_0, C_1\}$. Define $\sigma : S \to T$ by $s\sigma = [s]$, the largest integer less than or equal to s. For example, $(1/2)\sigma = 0$, $(3/2)\sigma = 1$, $(-3/4)\sigma = -1$, and $2\sigma = 2$. Define $\tau : T \to U$ by $(2n)\tau = C_0$ and $(2n + 1)\tau = C_1$ for $n \in I$. Then, for example, $(15/4)(\sigma \circ \tau) = [(15/4)\sigma]\tau = 3\tau = C_1$ and $(-29/3)(\sigma \circ \tau) = [(-29/3)\sigma]\tau = (-10)\tau = C_0$. Note that, in this case, $\tau \circ \sigma$ cannot be defined.

Given any nonempty set S we can consider the set, $M(S)$, of all 1–1 mappings of S onto itself and we have seen that we can consider the product, $\sigma \circ \tau$, of any two of these mappings. We now establish a series of three lemmas which will yield the result that $M(S)$ is indeed a group under the definition of the product of mappings.

Lemma 1. Suppose $\sigma : S \to T$ and $\tau : T \to U$ are mappings of S into T and T into U respectively and that both σ and τ are 1–1 and onto mappings. Then $\sigma \circ \tau$ is a 1–1 and onto mapping of S into U.

Proof. (1) Suppose $s_1, s_2 \in S$ with $s_1 \neq s_2$. Since σ is 1–1, $s_1\sigma \neq s_2\sigma$. But then, since τ is 1–1 and $s_1\sigma, s_2\sigma \in T$ it follows that $(s_1\sigma)\tau = s_1(\sigma \circ \tau)$ $\neq (s_2\sigma)\tau = s_2(\sigma \circ \tau)$. Hence $\sigma \circ \tau$ is 1–1.

(2) To establish that $\sigma \circ \tau$ is a mapping of S onto U we must show that, for any u in U, there exists an $s \in S$ such that $s(\sigma \circ \tau) = u$. Now since τ is an onto mapping of T into U, there exists $t \in T$ such that $t\tau = u$ and, since σ is an onto mapping of S into T, there exists $s \in S$ such that $s\sigma = t$. But then $s(\sigma \circ \tau) = (s\sigma)\tau = t\tau = u$ as desired.

Lemma 2. Suppose $\sigma : S \to T$, $\tau : T \to U$, and $\rho : U \to V$ are mappings of S into T, T into U, and U into V, respectively. Then $\sigma \circ (\tau \circ \rho) = (\sigma \circ \tau) \circ \rho$.

Proof. We must show (Definition 15) that, for all $s \in S$, we have $s[\sigma \circ (\tau \circ \rho)] = s[(\sigma \circ \tau) \circ \rho]$. But $s[\sigma \circ (\tau \circ \rho)] = (s\sigma)(\tau \circ \rho) = [(s\sigma)\tau]\rho = [s(\sigma \circ \tau)]\rho = s[(\sigma \circ \tau) \circ \rho]$ as desired.

Lemma 3. The mapping $\sigma : S \to T$ is a 1–1 mapping of S onto T if and only if there exists a mapping $\tau : T \to S$ such that $\sigma \circ \tau$ and $\tau \circ \sigma$ are the identity mappings on S and T respectively.

Proof. Suppose first that σ is a 1–1 mapping of S onto T. Then since σ is onto, given any $t \in T$, there exists $s \in S$ such that $s\sigma = t$; since σ is 1–1, this s is unique. We now define the mapping $\tau : T \to S$ by $t\tau = s$ if and only if $t = s\sigma$. This mapping, τ, is called the *inverse* of σ and we write $\tau = \sigma^{-1}$. We note that, for $s \in S$, $s(\sigma \circ \sigma^{-1}) = (s\sigma)\sigma^{-1} = t\sigma^{-1} = s$ so that $\sigma \circ \sigma^{-1}$ is the identity mapping of S onto S. Similarly, $t(\sigma^{-1} \circ \sigma) = (t\sigma^{-1})\sigma = s\sigma = t$ for every $t \in T$ so that $\sigma^{-1} \circ \sigma$ is the identity mapping of T onto T.

On the other hand, if $\sigma : S \to T$ is such that there exists a mapping $\tau : T \to S$ with the property that $\sigma \circ \tau$ and $\tau \circ \sigma$ are the identity mappings on S and T respectively, we can show that σ is a 1–1 mapping of S onto T. For σ is onto since, given $t \in T$, we have $t = t(\tau \circ \sigma) = (t\tau)\sigma$ where $t\tau \in S$. Furthermore, σ is 1–1 since if $s_1\sigma = s_2\sigma$ we have $(s_1\sigma)\tau = (s_2\sigma)\tau$ so that $s_1(\sigma \circ \tau) = s_2(\sigma \circ \tau)$ and $s_1 = s_2$.

Theorem 12. Let S be a nonempty set and $M(S)$ the set of all 1–1 mappings of S onto S. Then $M(S)$ is a group under the multiplication of mappings.

Proof. That the operation $\circ$ on these mappings is a binary operation on S was established by Lemma 1 and associativity follows by Lemma 2. The existence of an identity is obvious and the existence of inverses follows from Lemma 3.

Our primary interest here is with the case where S is a field or a group and the mappings are automorphisms.

Lemma 4. Let σ and τ be automorphisms of a field F. Then $\rho = \sigma \circ \tau$ is also an automorphism of F.

Before we prove this lemma let us consider an example of the product of two automorphisms.

Example 6. Let $F = \{a + b\sqrt{2} + ci + d\sqrt{2}i | a, b, c, d \in R\}$. We leave it to the student to show that:

(1) F is a field under the operations of addition and multiplication in R;

(2) If $(a + b\sqrt{2} + ci + d\sqrt{2}i)\sigma = a - b\sqrt{2} + ci - d\sqrt{2}i$, σ is an automorphism of F;

(3) If $(a + b\sqrt{2}i + ci + d\sqrt{2}i)\tau = a + b\sqrt{2} - ci - d\sqrt{2}i$, is an automorphism of F.

Then $\sigma \circ \tau$ is defined by

$$(a + b\sqrt{2} + ci + d\sqrt{2}i)(\sigma \circ \tau) = [(a + b\sqrt{2} + ci + d\sqrt{2}i)\sigma]\tau$$
$$= (a - b\sqrt{2} + ci - d\sqrt{2}i)\tau = a - b\sqrt{2} - ci + d\sqrt{2}i$$

and we again leave it to the student to show that $\sigma \circ \tau$ is an automorphism† of F.

Proof of Lemma.

(1) ρ is 1–1 and onto by Lemma 1.

(2) If $x, y \in F$, then $(x + y)\rho = (x + y)(\sigma \circ \tau) = [(x + y)\sigma]\tau = (x\sigma + y\sigma)\tau = (x\sigma)\tau + (y\sigma)\tau = x(\sigma \circ \tau) + y(\sigma \circ \tau) = x\rho + y\rho$ since σ and τ are automorphisms of F.

(3) If $x, y \in F$, then $(xy)\rho = (xy)(\sigma \circ \tau) = [(xy)\sigma]\tau = [(x\sigma)(y\sigma)]\tau = [(x\sigma)\tau][(y\sigma)\tau] = [x(\sigma \circ \tau)][y(\sigma \circ \tau)] = (x\rho)(y\rho)$ since σ and τ are automorphisms of F.

Theorem 13. The set of all the automorphisms of a field F forms a group. This group is called the *automorphism group* of the field.

Proof. This follows directly from Lemmas 1, 2, 3, and 4 and the observation that the identity mapping of F onto F is an automorphism of F.

It is obvious that if we considered only mappings that are preserved under addition we would also obtain a group of such correspondences. Since the elements of F form an Abelian group under addition, this means that the set of all automorphisms of an Abelian group themselves form a group. However, the student may check that nothing in the proof of Theorem 13 involves the use of commutativity of addition in F. Hence we have the following result.

† In fact, it may be shown that the only automorphisms of F are the identity, σ, τ, and $\sigma \circ \tau$.

Theorem 14. The set of all the automorphisms of a group forms a group called the *automorphism group* of the group.

The concept of an automorphism of a field may be used to prove, in a very neat fashion, an important theorem of elementary algebra.

Theorem 15. If $p(x) = a_0x^n + a_1x^{n-1} + \dots + a_n$ is a polynomial with real coefficients such that $p(a + bi) = 0$, where a and b are real numbers, then we also have $p(a - bi) = 0$.

Now the usual proof of this theorem is rather clumsy, involving a division of $p(x)$ by $[x - (a + bi)][x - (a - bi)]$ and an argument to show that the remainder upon this division is zero†.

On the other hand, if we first check that, if $(a + bi)\sigma = a - bi$ is an automorphism of C, we have, from

$$a_0(a + bi)^n + a_1(a + bi)^{n-1} + \dots + a_n = 0$$

that

$$[a_0(a + bi)^n + a_1(a + bi)^{n-1} + \dots + a_n]\sigma = 0\sigma = 0$$

$$(a_0\sigma)[(a + bi)\sigma]^n + (a_1\sigma)[(a + bi)\sigma]^{n-1} + \dots + a_n\sigma = 0$$

$$a_0(a - bi)^n + a_1(a - bi)^{n-1} + \dots + a_n = 0.$$

Exercises 7.7

1. Prove that, in Example 6, σ and τ are automorphisms of F.

2. Show why the following mappings on the set of real numbers, R^*, into R^* are not automorphisms of R^*. (Check *all* criteria: 1–1, onto, preservation under addition, and preservation under multiplication.)
   (a) $a\sigma = |a|$ (b) $a\sigma = \sqrt[3]{a}$
   (c) $a\sigma = 2a$ (d) $a\sigma = a^2$

3. Exhibit the multiplication table of the automorphisms of F in Example 6 (assuming there are only 4).

4. In F of Example 6 show that if $(a + b\sqrt{2} + ci + d\sqrt{2}i)\sigma = a - b\sqrt{2} - ci - d\sqrt{2}i$, σ is not an automorphism of F.

5. Do as in problem 4 for $(a + b\sqrt{2} + ci + d\sqrt{2}i)\tau = a + b\sqrt{2} - ci + d\sqrt{2}i$.

6. Let $F = \{a + b\sqrt[4]{3} + c\sqrt[4]{9} + d\sqrt[4]{27} + ei\sqrt[4]{3} + fi\sqrt[4]{9} + gi\sqrt[4]{27} + hi|a, b, c, d, e, f, g, h \in R\}$. Assume that F is a field and find as many automorphisms of F as you can. (There are eight.)

7. Construct a Cayley array for the automorphism group of problem 5, Exercises 7.6.

8. Show, by an example, that $\sigma \circ \tau$ can be an onto mapping even though σ and τ are not both onto mappings.

† See, for example, M. J. Weiss, *Higher Algebra for the Undergraduate*, Second Edition (1962), New York: John Wiley and Sons, p. 80.

9. Show, by an example, that $\sigma \circ \tau$ can be a 1–1 mapping even though σ and τ are not both 1–1 mappings.

***10.** Do as in problem 7 for the automorphism group of problem 6, Exercises 7.6.

***11.** Prove that if a set S has more than two elements, then $M(S)$ is not an Abelian group.

8. Cyclic Groups

Let us now return to the subject of groups by considering a special class of groups known as cyclic groups. We first make some preliminary definitions and remarks.

Definition 18. $a^0 = e$, $a^n = aa^{n-1}$ *for n a positive integer and a any element of a group G with identity e. Furthermore, we define* $a^{-n} = (a^{-1})^n$.

Thus $a^1 = aa^0 = ae = a$, $a^2 = aa^1 = aa$, $a^3 = a(aa)$ and so on.

For the proof of our next theorem we need the following two lemmas.

Lemma 5. Let G be a group and $a \in G$. Then $aa^n = a^n a$ for all $n \in N$.

Proof. The result is obviously true for $n = 1$. Now suppose it is true for $n = k : aa^k = a^k a$. Then

$$\begin{aligned} aa^{k+1} &= a(aa^k) && \text{Definition 18} \\ &= a(a^k a) && \text{Induction hypothesis} \\ &= (aa^k)a && \text{Associative law} \\ &= a^{k+1}a && \text{Definition 18.} \end{aligned}$$

Hence, by induction, $aa^n = a^n a$ for all $n \in N$.

Lemma 6. Let G be a group with identity e and $a \in G$. Then $(a^{-1})^n a^n = e$ for all $n \in N$.

Proof. The result is obviously true for $n = 1$. Now suppose it is true for $n = k : (a^{-1})^k a^k = e$. Then

$$\begin{aligned} (a^{-1})^{k+1}a^{k+1} &= [a^{-1}(a^{-1})^k][aa^k] && \text{Definition 18} \\ &= [(a^{-1})^k a^{-1}][aa^k] && \text{Lemma 5} \\ &= (a^{-1})^k[a^{-1}(aa^k)] && \text{Associative law} \\ &= (a^{-1})^k[(a^{-1}a)a^k] && \text{Associative law} \\ &= (a^{-1})^k(ea^k) && \text{Property of } a^{-1} \\ &= (a^{-1})^k a^k && \text{Property of } e \\ &= e && \text{Induction hypothesis} \end{aligned}$$

Hence, by induction, $(a^{-1})^n a^n = e$ for all $n \in N$.

Theorem 16. Let G be a group and $a \in G$. Then (1) $a^m a^n = a^{m+n}$ and (2) $(a^m)^n = a^{mn}$ for any integers m and n.

Proof. (1) Let $S_n = \{m | a^m a^n = a^{m+n}\}$ where $n = 0$ or $n \in N$. Then $0 \in S_n$ since $a^0 a^n = ea^n = a^n = a^{0+n}$. Also $1 \in S_n$ since $a^1 a^n = aa^n = a^{1+n}$. Suppose $k \in S_n$ for $k \in N$. Then $a^{k+1}a^n = (a^1 a^k)a^n = a(a^k a^n) = aa^{k+n} = a^{1+(k+n)} = a^{(k+1)+n}$. Thus if $k \in N$ it follows that $k + 1 \in N$. Hence, by induction, $N \subseteq S_n$ and we know that (1) is true for any non-negative integers m and n. Now if $m = -m'$ and $n = -n'$ where m' and $n' \in N$, we have

$$a^m a^n = a^{-m'}a^{-n'} = (a^{-1})^{m'}(a^{-1})^{n'} = (a^{-1})^{m'+n'}$$
$$= a^{-(m'+n')} = a^{(-m')+(-n')} = a^{m+n}$$

so that (1) is also true if m and n are both negative.

Now suppose that $n \in N$ and $m = -m'$ for $m' \in N$. If $n \geq m'$ we have

$$a^m a^n = a^{-m'}a^n = (a^{-1})^{m'}a^{m'}a^{n-m'}$$

since $n - m' \geq 0$. But $(a^{-1})^{m'}a^{m'} = e$ by Lemma 6 and hence $a^m a^n = a^{n-m'} = a^{n+(-m')} = a^{n+m}$. A similar proof handles the situation when $m' > n$ and when n is negative and m is positive.

(2) If $n = 0$, (2) certainly holds since $(a^m)^0 = e = a^0 = a^{m \cdot 0}$. Now let $S_m = \{n | (a^m)^n = a^{mn}\}$ for any integer m. Then $1 \in S_m$ since $(a^m)^1 = a^{m \cdot 1}$. Suppose $k \in S_m$. Then $(a^m)^{k+1} = a^m(a^m)^k = a^m a^{mk} = a^{m+mk} = a^{m(k+1)}$. Hence $k \in S_m$ implies $k + 1 \in S_m$ and, by induction, $N \subseteq S_m$. Now if $n = -n'$ where $n' \in N$ we have

$$(a^m)^n = (a^m)^{-n'} = [(a^m)^{-1}]^{n'}$$

But $(a^m)^{-1}$ is the inverse of a^m and, by (1), $a^m a^{-m} = a^0 = e$. Hence, since the inverse of any element is unique, we have $(a^m)^{-1} = a^{-m}$. Thus $(a^m)^n = (a^{-m})^{n'} = a^{(-m)n'}$ since we have shown that (2) holds for positive integers n. Finally, $(a^m)^n = a^{(-m)n'} = a^{(-m)(-n)} = a^{mn}$.

Definition 19. *Let G be a group with identity e and $a \in G$. If $a^m = e$ for some $m \in N$ but $a^k \neq e$ for $k \in N$ and $k < m$ we say that the element a has **order** m. On the other hand, if $m \in N$ implies $a^m \neq e$ we say that a is of **infinite order**.*

Example 1. In the group of symmetries of a square, the identity is of order 1 (as is the identity element of any group); R and R'' have order 4; and R', H, V, D, and D' all have order 2.

Example 2. In the additive group of integers, every element except the identity is of infinite order. (Note that, here, in the additive notation we replace a^n by na.)

We are now ready to define a cyclic group.

> ***Definition 20.*** *A group G is said to be* **cyclic** *if there is an element $a \in G$ such that $b \in G$ implies $b = a^n$ for some integer n. Such an element a is called a* **generator** *of G.*

Example 3. $G = \{e, a, a^2\}$ where $a^3 = e$ is a cyclic group of order 3 with generator a. (Another generator is a^2.)

Example 4. The additive group of integers is a cyclic group of infinite order with generator 1. (Another generator is -1.)

Theorem 17. (1) A cyclic group G with a generator of order n is isomorphic to the additive group of integers modulo n; and (2) a cyclic group G with a generator of infinite order is isomorphic to the additive group of integers.

Proof. (1) Let a be a generator of order n. If $n = 0$, then $G = \{a^0\} = \{e\} \cong \{C_0\}$. If $n \neq 0$, then $a^n = e$ but $a^k \neq e$ for $0 < k < n$. Furthermore, for $m \in N$ and $m > n$ we have $m = qn + r$ with $q, r \in N$ and $0 \leq r < n$. Then $a^m = a^{qn+r} = (a^n)^q a^r = e^q a^r = a^r$ so that, for positive exponents, we need consider only a^t for $0 < t < n$. Also we note that $aa^{n-1} = a^n = e$ so that $a^{-1} = a^{n-1}$. Hence, for any $k \in N$, $a^{-k} = (a^{-1})^k = (a^{n-1})^k = a^{k(n-1)}$ where $k(n-1) \geq 0$ so that we need consider only positive exponents, t, such that $0 < t < n$.

Now suppose that $a^r = a^s$ where $0 < r < n$, $0 < s < n$, and $s \geq r$. Then $e = a^{-r}a^r = a^{-r}a^s = a^{s-r}$ where $0 \leq s - r < n$. By definition of n, $s - r = 0$ so that $s = r$. Hence all of the elements $a, a^2, \ldots, a^{n-1}$, $a^n = e$ are distinct and $G = \{a, a^2, \ldots, a^{n-1}, e\}$ is of order n. If we now set up the 1–1 correspondence $a^s \in G \leftrightarrow C_s \in J_n$ for $s = 0, 1, \ldots, n-1$ we have $a^r a^s = a^{r+s} \leftrightarrow C_r + C_s = C_{r+s}$ where, in both uses of $r + s$, $r + s$ is calculated modulo n.

(2) Let a be a generator of G. Then, as in the proof of (1), we may show that $a^r \neq a^s$ for $r, s \in I$ unless $r = s$. Thus $G = \{a^s | s \in I\}$ and the correspondence $a^s \in G \leftrightarrow s \in I$ defines an isomorphism between G and the group of integers under addition.

Exercises 7.8

1. Do the nonzero elements of J_7 form a cyclic group under multiplication?
2. In J_8 do the elements C_1, C_3, C_5, and C_7 form a cyclic group under multiplication?

3. In J_9 do the elements C_1, C_2, C_4, C_5, C_7, and C_8 form a cyclic group under multiplication?
4. How many elements of the cyclic group of order 6 can be used as generators of the group?
5. Prove that an Abelian group of order 6 which contains an element of order 3 is a cyclic group.
6. If G is a cyclic group of order m generated by a, prove that a^k also generates G if and only if $(k, m) = 1$.
7. If G is a cyclic group of order m generated by a, find the order of any element, a^k, of G.
8. How many automorphisms has a cyclic group of order p? * One of order pq (p and q distinct primes)?

9. Subgroups

Definition 21. *A subset H of a group G is called a* ***subgroup*** *of G if H itself is a group under the group operation of G.*

Example. The subgroups of the symmetries of a square may be listed in the form of a "lattice" in which, for example, $\{I, D\}$ is a subgroup of $\{I, D, D', R'\}$ as well as of G itself.

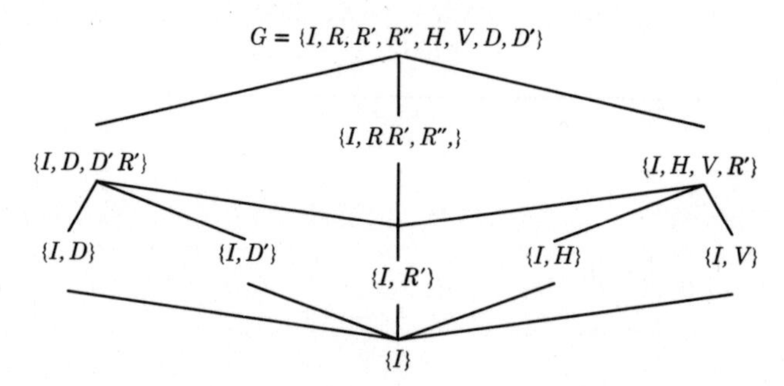

Note that every group G with identity e has at least two subgroups: G and $\{e\}$. (Of course if G is of order 1, $G = \{e\}$). By a *proper subgroup* of a group G we mean a subgroup other than G or $\{e\}$.

Theorem 18. A subset H of a group G is a subgroup of G if and only if
(1) $a \in H$ and $b \in H$ implies $ab \in H$;
(2) $a \in H$ implies $a^{-1} \in H$.
For finite groups (1) implies (2).

Proof. The necessity of the conditions is obvious from the definitions of group and subgroup.

On the other hand, conditions (1) and (2) assure closure, existence of identity ($aa^{-1} = e$), and existence of inverses; associativity follows since the elements of H are also elements of G.

Finally, we have $a, a^2, \ldots, a^n, \ldots \in G$ by (1). Thus if G is finite, not all of $a, a^2, \ldots, a^n, \ldots$ are distinct so that $a^r = a^s$ for some $r \neq s$. Suppose $r > s$; then $a^{r-s} = e$, $a^{r-s-1}a = e$, and a^{r-s-1} is the inverse of a. (If $r = s + 1$, $a^{r-s-1} = a^0 = e$ so that $ea = a$ and $a = e$.) That is, (1) implies (2) if G is finite.

In general, even for finite groups, it is not easy to determine all of the subgroups of a given group. For cyclic groups, however, we have the following theorem.

Theorem 19. Every subgroup of a cyclic group is cyclic.

Proof. Let G be a cyclic group generated by a and let H be a subgroup of G. If $a \in H$, $H = G$. Suppose $a^m \in H$ for $m > 1$, $m \in N$ but $a^k \notin H$ for $k \in N$ and $0 < k < m$. Then $a^{tm} \in H$ for every $t \in I$ since H is a subgroup of G. *If* $a^s \in H$ for $s > m$ we use the division algorithm to write $s = qm + r$, $0 \leq r < m$. Then $a^s = a^{qm+r} = a^{qm}a^r$. Thus $a^r = a^{-qm}a^s \in H$ so that, by definition of m, $r = 0$. Hence $s = qm$ and every element, a^s, of H is a power, $a^{mq} = (a^m)^q$, of a^m. Thus H is cyclic with generator a^m.

Corollary. If G is a finite cyclic group of order n with generator a and H is a subgroup of G, then H is generated by a^m for some $m \in N$ such that $m|n$ and the order of H is n/m.

We leave the proof of this corollary as an exercise for the student and remark that, in Chapter 9, we will show that, in general, the order of a subgroup of any group divides the order of the group.

Theorem 20. Every finite group G of order n is isomorphic to a group of permutations on n symbols.

Proof. Let $G = \{s_1, s_2, \ldots, s_n\}$ and define

$$S_j = \begin{pmatrix} s_1 & s_2 & \cdots & s_n \\ s_1s_j & s_2s_j & \cdots & s_ns_j \end{pmatrix} \qquad (j = 1, 2, \ldots, n)$$

We claim that S_j is then a permutation of the n symbols $s_1, s_2, \ldots, s_n$. Clearly this will be so if the s_ks_j $(k = 1, 2, \ldots, n)$ are distinct. But this is true since if $s_is_j = s_ks_j$, $s_i = s_k$ by the cancellation law for groups.

Now we observe that $\{S_1, S_2, \ldots, S_n\}$ forms a subgroup of the symmetric group on n symbols. For, by Theorem 18, we need only establish closure and this follows since

$$S_jS_k = \begin{pmatrix} s_1 & s_2 & \cdots & s_n \\ s_1s_j & s_2s_j & \cdots & s_ns_j \end{pmatrix}\begin{pmatrix} s_1s_j & s_2s_j & \cdots & s_ns_j \\ (s_1s_j)s_k & (s_2s_j)s_k & \cdots & (s_ns_j)s_k \end{pmatrix}$$

$$= \begin{pmatrix} s_1 & s_2 & \cdots & s_n \\ (s_1s_j)s_k & (s_2s_j)s_k & \cdots & (s_ns_j)s_k \end{pmatrix}$$

and $(s_is_j)s_k = s_i(s_js_k) = s_is_m$ where $s_m = s_js_k \in G$. Thus $S_jS_k = S_m$.

Our isomorphism, of course, is provided by the correspondence $s_j \leftrightarrow S_j$ since if $s_js_k = s_m$, then

$$s_m = s_js_k \leftrightarrow S_m = S_jS_k$$

There may well be, of course, other permutation groups isomorphic to any given group G. The one obtained in the fashion just described is called the ***regular permutation*** group isomorphic to G.

Example 1. Consider the group $\{s_1, s_2, s_3, s_4\}$ with Cayley array

	s_1	s_2	s_3	s_4
s_1	s_1	s_2	s_3	s_4
s_2	s_2	s_1	s_4	s_3
s_3	s_3	s_4	s_1	s_2
s_4	s_4	s_3	s_2	s_1

Then

$$s_1 \leftrightarrow S_1 = \begin{pmatrix} s_1 & s_2 & s_3 & s_4 \\ s_1s_1 & s_2s_1 & s_3s_1 & s_4s_1 \end{pmatrix} = \begin{pmatrix} s_1 & s_2 & s_3 & s_4 \\ s_1 & s_2 & s_3 & s_4 \end{pmatrix}$$

$$s_2 \leftrightarrow S_2 = \begin{pmatrix} s_1 & s_2 & s_3 & s_4 \\ s_1s_2 & s_2s_2 & s_3s_2 & s_4s_2 \end{pmatrix} = \begin{pmatrix} s_1 & s_2 & s_3 & s_4 \\ s_2 & s_1 & s_4 & s_3 \end{pmatrix}$$

$$s_3 \leftrightarrow S_3 = \begin{pmatrix} s_1 & s_2 & s_3 & s_4 \\ s_1s_3 & s_2s_3 & s_3s_3 & s_4s_3 \end{pmatrix} = \begin{pmatrix} s_1 & s_2 & s_3 & s_4 \\ s_3 & s_4 & s_1 & s_2 \end{pmatrix}$$

$$s_4 \leftrightarrow S_4 = \begin{pmatrix} s_1 & s_2 & s_3 & s_4 \\ s_1s_4 & s_2s_4 & s_3s_4 & s_4s_4 \end{pmatrix} = \begin{pmatrix} s_1 & s_2 & s_3 & s_4 \\ s_4 & s_3 & s_2 & s_1 \end{pmatrix}$$

or, more simply, $s_1 \leftrightarrow (1)(2)(3)(4)$, $s_2 \leftrightarrow (1\ 2)\ (3\ 4)$, $s_3 \leftrightarrow (1\ 3)(2\ 4)$, $s_4 \leftrightarrow (1\ 4)(2\ 3)$.

Exercises 7.9

1. Exhibit the proper subgroups of the additive group J_{12} of residue classes modulo 12.
2. Find all of the subgroups of the cyclic group of order 12.
3. Let S and T be two subgroups of a group G. Prove that $S \cap T$ is a subgroup of G.
4. Let G be a group and $S = \{x | x \in G \text{ and } ax = xa \text{ for all } a \in G\}$. Prove that S is a subgroup of G. (It is called the *center* of G.)
5. Find the regular permutation group isomorphic to the cyclic group of order 5.
6. Do as in problem 5 for the cyclic group of order 6.
7. Do as in problem 5 for the *octic* group defined as $\{s_1, s_2, \ldots, s_8\}$ where $s_1 = (1)(2)(3)(4)$, $s_2 = (1234)$, $s_3 = (13)(24)$, $s_4 = (1432)$, $s_5 = (13)$, $s_6 = (24)$, $s_7 = (12)(34)$, and $s_8 = (14)(23)$.
8. Prove the corollary to Theorem 19.

*9. Prove that if a and b are elements of a group G much that $ab \neq ba$, then the five elements a, b, ab, ba, and e (the identity of G) are distinct. Moreover, prove that one of a^2 and aba is distinct from all these five. Use these results to prove that a noncommutative group has at least six elements.

*10. Prove that a group that has only finitely many subgroups must be a finite group.

PROBLEMS FOR FURTHER INVESTIGATION

1. In problem 3 of Exercises 7.3 you were asked to prove that a group G was Abelian if and only if $(ab)^2 = a^2b^2$ for all $a, b, \in G$. What, if anything, can you say about a group in which $(ab)^n = a^nb^n$ for all $a, b \in G$ where n is some fixed natural number other than 2? [Z1]
2. Consider the problem of determining all non-isomorphic groups of various orders. In particular consider the case when the order is ≤ 6. What about the special case of Abelian groups? [B4], [C2]
3. Find conditions on a finite cyclic group such that its automorphism group is cyclic. If the group G is cyclic and its automorphism group, G_1, is cyclic, consider the "chain" of groups, $G_2, G_3, \ldots$ where G_2 is the automorphism group of G_1, G_3 is the automorphism group of G_2 and so on. What are the conditions on G if $G_1, G_2, G_3, \ldots$ are all cyclic? [D7]
4. The postulates defining a group are, of course, not unique. Problem 6 of Exercises 7.3 gives an alternative list of postulates. Consider the problem of other formulations of the postulates for a group. [S1],]S2], [W3]
5. Consider the possibility of dropping one or more of the postulates for a group. What theorems, if any, can you prove concerning such a system? In particular, suppose that only closure and the associative law are assumed or that, on the other hand, all of the group postulates are satisfied except the associative law. [A2], [M5]
6. From the integral domain of the integers we constructed the field of rational numbers in Chapter 6. Consider the problem of constructing, in a similar fashion, a field (called a *quotient field*) from any integral domain. [B4]

REFERENCES FOR FURTHER READING

[B4], [C2], [D1], [E1], [H6], [K4], [Z1].

8

The Real Numbers

We return now to the development of our number system. In our previous discussions we motivated our development through the consideration of equations: If a and b are natural numbers with $a \geq b$, there is no solution in natural numbers for the equation $x + a = b$; hence the integers were developed. Similarly, if a and b are integers such that $a \nmid b$, there is no solution in integers for the equation $ax = b$; hence the rational numbers were developed.

Now consider the equation $x^2 - 2 = 0$. Is there a rational number, a/b, a, $b \in I$, such that $(a/b)^2 - 2 = 0$? As the student is probably well aware, the answer is "no". A proof of this assertion is as follows: If a and b have a common factor m we have $a/b = (ma_1)/(mb_1) = a_1/b_1$. Thus if we assume that there exists a rational number a/b such that $(a/b)^2 - 2 = 0$, then there also exists a rational number a_1/b_1, such that $(a_1/b_1)^2 - 2 = 0$ and $(a_1, b_1) = 1$. Hence we may assume to begin with that $(a, b) = 1$.

Now since $(a/b)^2 - 2 = 0$, $a^2 = 2b^2$ and a^2 is even. But if $2n + 1$ is any odd integer, its square is $(2n + 1)^2 = 4n^2 + 4n + 1 = 2(2n^2 + 2n) + 1$. Thus the square of any odd integer is odd and hence the fact that a^2 is even implies that a is even. We conclude that $a = 2m$ so that $a^2 = (2m)^2 = 4m^2 = 2b^2$. Hence $b^2 = 2m^2$ so that b is also even contrary to our hypothesis that $(a, b) = 1$. Thus our assumption that there exists a rational number a/b such that $(a/b)^2 - 2 = 0$ is false.

We are thus led to the construction of additional numbers to enable us to solve such equations as $x^2 - 2 = 0$ and, more generally, to fill in the "gaps" in the real number line. The situation, however, is not nearly as simple as in our development of the integers and the rational numbers since it turns out that we cannot take pairs of rational numbers (as the student might expect). We shall not go into the reasons for this now, but will consider this aspect of the situation later.

1. Rational Numbers as Decimals

The motivation for our treatment of the real numbers comes from the fact that all rational numbers may be expressed as certain kinds of decimals.

Thus some rational numbers can be expressed as *terminating decimals*:

$$2 = 2.0, \qquad \tfrac{1}{4} = 0.25, \qquad \tfrac{1}{8} = 0.125$$

and so on and all other rational numbers can be expressed as *repeating decimals*:

$$\tfrac{1}{3} = 0.3333\ldots, \qquad \tfrac{2}{7} = 0.285714285714\ldots, \qquad 3\tfrac{5}{6} = 3.83333\ldots.$$

The reason for this is quite simple. In expressing a/b ($a, b \in N$) as a decimal by dividing b into a, our successive remainders in the usual division process are always less than b. Thus the only possibilities for remainders are $0, 1, \ldots, b - 1$ and hence if the division is not exact the remainders must eventually repeat.

Conversely, if we have given any terminating decimal it is obvious that we have a rational number: $0.25 = 25/100$, $0.125 = 125/1000$ etc. Furthermore, as we will prove later, every repeating decimal also represents a rational number. Since we can consider any terminating decimal as a repeating decimal with 0's (e.g. $2 = 2.0000\ldots$) we need only consider repeating decimals when we are discussing the rational numbers.

The situation then is this: we have a set of repeating decimals that we associate with the rational numbers. Nonrepeating decimals, then, such as

$$0.1010010001\ldots$$

(where we add, after each 1, an additional 0 as a digit) should represent the irrational numbers.

All of this, however, is very imprecise since we have no formal definition of a decimal, equality of decimals, and addition and multiplication of decimals. The rest of this chapter is devoted to making precise the intuitive notions that we have just sketched.

Exercises 8.1

1. Express the following rational numbers as terminating or repeating decimals:
(a) 2/9 (b) 3/11
(c) 3/16 (d) 4/13

2. Prove that the following are all irrational numbers:
(a) $\sqrt{3}$ (b) $\sqrt[3]{2}$
(c) $\log_{10} 3$ (d) $\sqrt{2} + \sqrt{3}$

2. Sequences

The student has met many examples of sequences in his previous work in mathematics; for example, arithmetic sequences such as 2, 5, 8, 11,

$\ldots, 3k - 1, \ldots$; geometric sequences such as $2, 6, 18, \ldots, 2 \cdot 3^{k-1}, \ldots$; and the harmonic sequence $1, \frac{1}{2}, \frac{1}{3}, \ldots, 1/k, \ldots$. We will consider here sequences $a_1, a_2, \ldots, a_k, \ldots$ of rational numbers; that is, $a_1, a_2, \ldots, a_k, \ldots$ will be rational numbers. For brevity we write $[a_k]$ for the sequence $a_1, a_2, \ldots, a_k, \ldots$.

Definition 1. $[a_k] = [b_k]$ *if and only if* $a_i = b_i$ *for all* $i \in N$.

Definition 2. $[a_k] + [b_k] = [a_k + b_k]$.

Definition 3. $[a_k] \times [b_k] = [a_k][b_k] = [a_k b_k]$.

Two remarks are in order on these definitions:

(1) Our equality for sequences is identity and hence we may use the symbol "=" without any qualification.

(2) In Definition 2 we should really use $\oplus$, say, for the $+$ on the left hand side of the equality to distinguish the (defined) addition of sequences from the addition of rational numbers in $[a_k + b_k]$. The student, however, should have no difficulty in distinguishing these two uses of $+$. A similar remark, of course, applies to Definition 3.

It should be clear that some sequences are associated in a natural fashion with real numbers: the sequence 0.3, 0.33, 0.333, ... is associated with $\frac{1}{3}$; 1.4, 1.41, 1.414,... with $\sqrt{2}$ etc. Before this can be done, however, we must "outlaw" sequences whose terms become larger without bound (as $1, 2, \ldots, k, \ldots$) and those whose terms fluctuate wildly (as $0, 1, 0, 1, 0, \ldots, (1/2) + (-1)^k(1/2), \ldots$). We will do this in the next section, but right now let us make two further definitions and state a basic theorem concerning the arithmetic of sequences.

Definition 4. *If* $[a_k]$ *is a sequence,* $-[a_k]$ *is defined to be the sequence* $[-a_k]$.

Definition 5. *The sequence* $[a_k]$ *with* $a_k = 1$ *for all* $k \in N$ *is called the* ***unit*** *sequence and the sequence* $[a_k]$ *with* $a_k = 0$ *for all* $k \in N$ *is called the* ***zero*** *sequence. We write* $[1]$ *for the unit sequence and* $[0]$ *for the zero sequence. In general, if* $a_k = r \in R$ *for all* $k \in N$*, we write* $[a_k] = [r]$.

We leave as an exercise for the student to prove, using Definitions 1–5 and the properties of the rational numbers, the following theorem.

Theorem 1. Let $S = \{[a_k] | a_k \in R\}$. Then S is a commutative ring with a unity under the operations of addition and multiplication as given in Definitions 2 and 3.

Exercises 8.2

1. Prove Theorem 1.
2. Prove that $S = \{[a_k] | a_k \in R\}$ is not an integral domain by exhibiting divisors of zero.

3. Cauchy Sequences

Our way of insuring that a sequence will not fluctuate unduly or be unbounded is to demand that it satisfy the *Cauchy condition.*

Definition 6. *Let R^+ be the set of positive rational numbers. If for any $\alpha \in R^+$ there exists a natural number $\phi(\alpha)$ such that when $n \geq \phi(\alpha)$ and $m \geq \phi(\alpha)$, then $|a_n - a_m| < \alpha$, we call $[a_k]$ a **Cauchy sequence.** The condition that $|a_n - a_m| < \alpha$ is called the **Cauchy condition.***

Example 1. Suppose $a_k = 1 - 1/2^k$. Then consider $|a_n - a_m|$ where, since we are using absolute values, we may take $n \geq m$ so that $1/2^m \geq 1/2^n$. We have

$$|a_n - a_m| = \left|\left(1 - \frac{1}{2^n}\right) - \left(1 - \frac{1}{2^m}\right)\right| = \left|\frac{1}{2^m} - \frac{1}{2^n}\right| < \frac{1}{2^m}$$

Thus to have $|a_n - a_m| < \alpha$ we need only take m such that $1/2^m \leq \alpha$ or, equivalently, choose m such that $2^m \geq 1/\alpha$. Clearly such a choice of $m \in N$ is always possible and hence $[a_k]$ is a Cauchy sequence. (A formal proof, using the Archimedean law (Theorem 6.13), that this choice is possible is left as an exercise for the student.)

Example 2. Suppose that our sequence is 1, 0, 1, 0, 1, ... where $a_k = 1$ if k is odd and $a_k = 0$ if k is even. Then no matter how large n is we have $|a_{n+1} - a_n| = 1$ and hence $[a_k]$ is not a Cauchy sequence.

Example 3. Suppose our sequence is $1, 2, \ldots, k, \ldots$. Again $|a_{n+1} - a_n| = 1$ and $[a_k]$ is not a Cauchy sequence.

It is intuitively plausible that the terms of a Cauchy sequence cannot get arbitrarily large since the later terms differ by an arbitrarily small amount. We phrase this fact more formally in our next theorem.

Theorem 2. If $[a_k]$ is a Cauchy sequence, then there exists a $B \in R^+$ such that $|a_k| < B$ for all $k \in N$. (Such a number B is called an *upper bound* for the sequence.)

Proof. By the properties of absolute value (Theorem 5.9) we have

$$|a_k| = |a_r + (a_k - a_r)| \leq |a_r| + |a_k - a_r|$$

Now since $[a_k]$ is a Cauchy sequence there exists $\phi(1) \in N$ such that when $k, r \geq \phi(1)$, $|a_k - a_r| < 1$. Hence for all $k, r \geq \phi(1)$ we have

$$|a_k| < |a_r| + 1$$

Now pick any $r \geq \phi(1)$; say $r = \phi(1)$. Then, for $k \geq \phi(1)$ we have $|a_k| < |a_r| + 1 = c$ where $c \in R^+$ is a constant. Thus $|\alpha_k| < c$ for all $k \geq \phi(1)$. To take care of $|a_k|$ for $k < \phi(1)$ we consider the finite set of positive rational numbers $\{|a_1|, |a_2|, \ldots, |a_{\phi(1)}|, c\}$ and let B be any rational number such that $B > |a_i|$ for $i = 1, 2, \ldots, \phi(1)$ and also $B > c$ (e.g., take $B = |a_1| + |a_2| + \ldots + |a_{\phi(1)}| + c$). Then $|a_k| < B$ for all $k \in N$.

NOTE. The choice of $\alpha = 1$ in $\phi(\alpha)$ above is purely arbitrary. Any $\alpha \in R^+$ would do.

Example 4. Suppose $a_1 = 20, a_2 = -3, a_3 = 4, a_5 = 1 + \frac{1}{2}, a_6 = 1 + \frac{1}{4}, \ldots, a_k = 1 + 1/2^{k-4}$ for $k > 4$. Then for $k, r \geq 5$, $|a_k - a_r| < 1$. Thus for all $k \geq 5$, $|a_k| < |a_5| + 1 = 5/2$. Now let $B = |20| + |-3| + |4| + 5/2 = 29\frac{1}{2}$. Clearly $|a_k| < B$ for all $k \in N$. (It is also clear that smaller upper bounds exist; e.g. 20.)

Theorem 3. The sum of two Cauchy sequences is a Cauchy sequence and the product of two Cauchy sequences is a Cauchy sequence.

Proof. Let $[a_k]$ and $[b_k]$ be Cauchy sequences. Then $[a_k] + [b_k] = [a_k + b_k]$. Consider

$$\begin{aligned} &|(a_n + b_n) - (a_m + b_m)| \\ &\qquad = |(a_n - a_m) + (b_n - b_m)| \leq |a_n - a_m| + |b_n - b_m| \end{aligned}$$

by the properties of absolute value. Now given any $\alpha/2 \in R^+$ we can find $\phi_1(\alpha/2) \in N$ such that when $n, m \geq \phi_1(\alpha/2)$, $|a_n - a_m| < \alpha/2$. Similarly, we can find $\phi_2(\alpha/2) \in N$ such that when $n, m \geq \phi_2(\alpha/2)$, $|b_n - b_m| < \alpha/2$.

Now here and elsewhere we will find it convenient to write *max* (a, b) to mean the larger of a and b if $a \neq b$ and a if $a = b$. Thus here we let $\phi(\alpha) = max\,(\phi_1(\alpha/2), \phi_2(\alpha/2))$ and then have

$$|(a_n + b_n) - (a_m + b_m)| \leq |a_n - a_m| + |b_n - b_m| < \alpha/2 + \alpha/2 = \alpha$$

for $n, m \geq \phi(\alpha)$. Hence $[a_k + b_k]$ is a Cauchy sequence.

Now consider $[a_k][b_k] = [a_k b_k]$. We have

$$\begin{aligned} |a_n b_n - a_m b_m| &= |a_n(b_n - b_m) + b_m(a_n - a_m)| \leq |a_n(b_n - b_m)| \\ &\quad + |b_m(a_n - a_m)| = |a_n||b_n - b_m| + |b_m||a_n - a_m| \end{aligned}$$

by the properties of absolute value. Now from Theorem 2 we know that there exist $B_1, B_2 \in R^+$ such that, for all $n, m \in N$, $|a_n| < B_1$ and $|b_m| < B_2$. Hence

$$|a_n b_n - a_m b_m| < B_1|b_n - b_m| + B_2|a_n - a_m|$$

Now we use the fact that $[a_k]$ and $[b_k]$ are Cauchy sequences to find $\phi_1(\alpha/2B_1) \in N$ such that for $n, m \geq \phi_1(\alpha/2B_1)$, $|b_n - b_m| < \alpha/2B_1$ and $\phi_2(\alpha/2B_2) \in N$ such that for $n, m \geq \phi_2(\alpha/2B_2)$, $|a_n - a_m| < \alpha/2B_2$. Take $\phi(\alpha) = max\,(\phi_1(\alpha/2B_1), \phi_2(\alpha/2B_2))$ and have for $n, m \geq \phi(\alpha)$,

$$|a_n b_n - a_m b_m| < B_1(\alpha/2B_1) + B_2(\alpha/2B_2) = \alpha$$

as desired.

Since the sequence $[a_k] = [0]$ and the sequence $[b_k] = [1]$ are obviously Cauchy sequences, we see that Theorems 1 and 3 yield the following result.

Theorem 4. Let $S = \{[a_k] | [a_k] \text{ is a Cauchy sequence}\}$. Then S is a commutative ring with a unity element under the operations of addition and multiplication as given in Definitions 2 and 3.

Exercises 8.3

1. Prove, using the Achimedean law, that for any $\alpha \in R^+$, there exists $m \in N$ such that $2^m \geq 1/\alpha$.
2. Prove that $S = \{[a_k] | [a_k] \text{ is a Cauchy sequence}\}$ is not an integral domain by exhibiting divisors of zero.
3. Prove that if B is an upper bound for a sequence $[a_k]$ and $A > B$, then A is also an upper bound for the sequence $[a_k]$.
4. A sequence $[a_k]$ is said to be *bounded* if there exists $B \in R^*$ such that for all $k \in N$, $|a_k| < B$. Prove that the sum and product of bounded sequences are bounded sequences.

4. Null Sequences

Let us recall the motivation behind our discussion to date. We are trying to associate sequences with our intuitive notion of a real number. Our problem is that there is a natural association of more than one sequence with any real number. Thus, for example, the sequences

$$[1 - 1/k], \qquad [1 - 1/2^k], \qquad \text{and} \qquad [1 - 1/k^2]$$

are all naturally associated with the number 1. Our basic device for handling this problem is that of a *null sequence*.

Definition 7. *$[a_k]$ is a **null sequence** if and only if for every $\alpha \in R^+$ there exists $\phi(\alpha) \in N$ such that $n \geq \phi(\alpha)$ implies $|a_n| < \alpha$.*

Example 1. $[1/k]$, $[1/2^k]$, and $[1/k^2]$ are all null sequences.

We leave to the student the proof of the following two theorems.

Theorem 5. Every null sequence is a Cauchy sequence.

Theorem 6. Let $[a_k]$ and $[b_k]$ be null sequences and $[c_k]$ be any Cauchy sequence. Then

(1) $[a_k] + [b_k]$ is a null sequence;
(2) $[a_k] - [b_k]$ is a null sequence;
(3) $[a_k][c_k]$ is a null sequence.

We are now ready to define a relation on the set of Cauchy sequences that will make precise the idea that two different Cauchy sequences define the same real number.

Definition 8. *Two Cauchy sequences $[a_k]$ and $[b_k]$ will be said to be* ***equivalent*** *if and only if $[a_k] - [b_k] = [a_k - b_k]$ is a null sequence. We will write $[a_k] \sim [b_k]$ if $[a_k]$ is equivalent to $[b_k]$.*

Example 2. The sequences $[1 - 1/k]$ and $[1 - 1/2^k]$ are equivalent since $[(1 - 1/k) - (1 - 1/2^k)] = [1/2^k - 1/k]$ is a null sequence.

Theorem 7. Let $S = \{[a_k] | [a_k] \text{ a Cauchy sequence}\}$. Then $\sim$ is an equivalence relation on S.

Proof. (1) $[a_k] \sim [a_k]$ since $[a_k - a_k] = [0]$ is certainly a null sequence

(2) If $[a_k] \sim [b_k]$, then $[a_k - b_k]$ is a null sequence. But then $[b_k - a_k]$. $= [-(a_k - b_k)]$ is also a null sequence and hence $[b_k] \sim [a_k]$.

(3) If $[a_k] \sim [b_k]$ and $[b_k] \sim [c_k]$, then $[a_k - b_k]$ and $[b_k - c_k]$ are null sequences. Hence, by Theorem 6, $[a_k - b_k] + [b_k - c_k] = [(a_k - b_k) + (b_k - c_k)] = [a_k - c_k]$ is a null sequence and $[a_k] \sim [c_k]$.

Exercises 8.4

1. Prove Theorem 5.
2. Prove Theorem 6.
3. Prove that the set of all null sequences is a commutative ring without a multiplicative identity.

5. The Real Numbers

Since $\sim$ is an equivalence relation on the set of all Cauchy sequences, it partitions the set of all Cauchy sequences into classes of equivalent sequences. Thus, for example, we have the class

$$\{[a_k] | [a_k] \sim [1]\}$$

and the class

$$\left\{[a_k] \middle| [a_k] \sim \left[\sum_{i=1}^{k} \frac{i}{i^2+1}\right]\right\}$$

Definition 9. *The set of all equivalence classes of* $\{[a_k] | [a_k]$ *a Cauchy sequence*$\}$ *is the set of* ***real numbers,*** R^*.

We now need to define addition and multiplication for these classes.

Definition 10. *Let* $A = \{[x_k] | [x_k] \sim [a_k]\}$ *and* $B = \{[y_k] | [y_k] \sim [b_k]\}$. *Then*

$$A + B = \{[z_k] | [z_k] \sim [a_k] + [b_k]\}$$

and

$$A \times B = AB = \{[z_k] | [z_k] \sim [a_k][b_k]\}$$

Here, of course, we should write $A \oplus B$ and $A \otimes B$ to distinguish the newly defined addition of classes from the previously defined addition and multiplication of sequences. At this point, however, the student can certainly keep this distinction in mind without the need of a special notation.

As usual, we need to prove that our definitions are independent of our choice of representative of our equivalence classes.

Theorem 8. If $\{[x_k] | [x_k] \sim [a_k]\} = \{[x_k] | [x_k] \sim [a'_k]\}$ and $\{[y_k] | [y_k] \sim [b_k]\} = \{[y_k] | [y_k] \sim [b'_k]\}$, then

(1) $\{[z_k] | [z_k] \sim [a_k] + [b_k]\} = \{[z_k] | [z_k] \sim [a'_k] + [b'_k]\}$

and

(2) $\{[z_k] | [z_k] \sim [a_k][b_k]\} = \{[z_k] | [z_k] \sim [a'_k][b'_k]\}$.

Proof. Because of the properties of $\sim$ and the definitions of addition and multiplication of equivalence classes we need only prove that if $[a_k] \sim [a'_k]$ and $[b_k] \sim [b'_k]$, then $[a_k] + [b_k] \sim [a_k] + [b'_k]$ and $[a_k][b_k] \sim [a'_k][b'_k]$. Now $([a_k] + [b_k]) - ([a'_k] + [b'_k]) = [a_k - a'_k] + [b_k - b'_k]$. Furthermore, $[a_k - a'_k]$ and $[b_k - b'_k]$ are null sequences since $[a_k] \sim [a'_k]$ and $[b_k] \sim [b'_k]$. Invoking Theorem 6 we conclude that $([a_k] + [b_k]) - ([a'_k] + [b'_k])$ is a null sequence and thus have established (1).

(For (2) we write

$$[a_k][b_k] - [a'_k][b'_k] = [a_k]([b_k] - [b'_k]) + ([a_k] - [a'_k])[b'_k]$$

Now $[b_k] - [b'_k]$ is a null sequence since $[b_k] \sim [b'_k]$ and $[a_k] - [a'_k]$ is a null sequence since $[a_k] \sim [a'_k]$. We now invoke Theorem 6 to conclude that $[a_k][b_k] - [a'_k][b'_k]$ is a null sequence and hence that $[a_k][b_k] \sim [a'_k][b'_k]$.

Because we have defined the operations of addition and multiplication in R^* in terms of representatives of the equivalence classes that constitute R^*, we may now invoke Theorem 4 to obtain

Theorem 9. R^* is a commutative ring with additive identity $\bar{0} = \{[x_k] | [x_k] \sim [0]\}$ and multiplicative identity $\bar{1} = \{[x_k] | [x_k] \sim [1[\}$. If $X = \{[x_k] | [x_k] \sim [a_k]\}$, then $-X = \{[x_k] | [x_k] \sim [-a_k]\}$.

Exercises 8.5

1. Discuss the difference between the natural number 2, the integer 2, the rational number 2, and the real number 2.
2. Give the details of the proof of Theorem 9.
3. Consider the example that you previously used to show that the set of all Cauchy sequences was not an integral domain. What "happens" to this example here? (i.e., why cannot it be used to show that R^* is not an integral domain?)

6. The Rational Numbers as Real Numbers

Ultimately we will show that R^* is an ordered field but let us first consider how the rational numbers appear as real numbers. Among the elements of R^* are those classes $\{[x_k] | [x_k] \sim [r]\}$ where $r \in R$. Let us, for brevity write $\{[x_k] | [x_k] \sim [r]\} = \{[r]\}$.

Theorem 10. The rational numbers are isomorphic to the set of elements $\{[r]\}$ of R^* via the correspondence

$$r \in R \leftrightarrow \{[r]\} \in R^*$$

Proof. This follows immediately from the fact that if $r_1, r_2 \in R$, then $\{[r_1]\} + \{[r_2]\} = \{[r_1] + [r_2]\} = \{[r_1 + r_2]\}$ and $\{[r_1]\} \times \{[r_2]\} = \{[r_1][r_2]\} = \{[r_1 r_2]\}$.

At this point let us return to the statement made in Section 1 that every repeating decimal represents a rational number. First of all, let us note that every repeating decimal can be represented as a rational number plus a sum of terms in a geometric progression. Thus, for example,

$$0.3333 \cdots = 0.3 + (0.3)(\tfrac{1}{10}) + (0.3)(\tfrac{1}{10})^2 + \cdots + (0.3)(\tfrac{1}{10})^{n-1} + \cdots,$$

$$7.4393939 \cdots = 7.4 + 0.039 + (0.039)(\tfrac{1}{100}) + \cdots + (0.039)(\tfrac{1}{100})^{n-2} + \cdots,$$

$$1.237237 \cdots = 1 + 0.237 + (0.237)(\tfrac{1}{1000}) + \cdots + (0.237)(\tfrac{1}{1000})^{n-2} + \cdots.$$

In general we have

$$r_1 + r_2 +_2(\tfrac{1}{10^s}) + r_2(\tfrac{1}{10^s})^2 + \cdots + r_2(\tfrac{1}{10^s})^{n-1} + \cdots$$

where $r_1, r_2 \in R$ and $s \in N$.

Consider now the sequence $[a_k]$ where

$$a_k = r_1 + r_2 + r_2(\tfrac{1}{10^s}) + r_2(\tfrac{1}{10^s})^2 + \ldots + r_2(\tfrac{1}{10^s})^{k-1}$$

We will show that

$$[a_k] \sim \left[r_1 + \frac{r_2}{1 - (\frac{1}{10})^s}\right] = [r]$$

and thus that

$$\{[x_n] | [x_n] \sim [a_k]\} = \{[x_n] | [x_n] \sim [r]\} = \{[r]\} \leftrightarrow r \in R$$

in the sense of Theorem 10. For example, for 0.3333 ... we have $r_1 = 0$, $r_2 = 0.3$, $s = 1$ so that

$$r = r_1 + \frac{r_2}{1 - (\frac{1}{10})^s} = 0 + \frac{.3}{1 - \frac{1}{10}} = \frac{3}{10} \div \frac{9}{10} = \frac{1}{3}$$

and for 7.43939 ... we have $r_1 = 7.4$, $r_2 = 0.039$, $s = 2$ so that

$$r = r_1 + \frac{r_2}{1 - (\frac{1}{10})^s} = 7.4 + \frac{0.039}{1 - (\frac{1}{10})^2} = \frac{74}{10} + \left(\frac{39}{1000} \div \frac{99}{100}\right)$$
$$= \frac{37}{5} + \frac{13}{330} = \frac{12{,}275}{1650} = \frac{491}{66}$$

Now from the theory of geometric progressions† we know that if $S = b + bc + bc^2 + \cdots + bc^{n-1}$, then

$$S = b\frac{c^n - 1}{c - 1}$$

Applying this formula to $a_k - r_1$ with $b = r_2$, $c = 1/10^s$, and $n = k$ we have

$$a_k = r_1 + r_2\frac{(\frac{1}{10})^{sk} - 1}{(\frac{1}{10})^s - 1}$$

Thus to show that $[a_k] \sim [r]$ we need only show that

$$[a_k - r] = \left[\left(r_1 + r_2\frac{(\frac{1}{10})^{sk} - 1}{(\frac{1}{10})^s - 1}\right) - \left(r_1 + \frac{r_2}{1 - (\frac{1}{10})^s}\right)\right]$$
$$= \left[\frac{r_2}{(\frac{1}{10})^s - 1} \cdot (\tfrac{1}{10})^{sk}\right]$$

† See any college algebra textbook.

is a null sequence. That is, given any $\alpha \in R^+$ we want to be able to choose $\phi(\alpha) \in N$ such that, when $n \geq \phi(\alpha)$,

$$\left| \frac{r_2}{(\frac{1}{10})^s - 1} \cdot \left(\frac{1}{10}\right)^{sn} \right| = \frac{r_2}{1 - (\frac{1}{10})^s} \cdot \left(\frac{1}{10}\right)^{sn} < \alpha$$

Thus we want

$$(\tfrac{1}{10}{}^s)^n < \alpha \frac{1 - (\frac{1}{10})^s}{r_2}$$

or

$$(10^s)^n > \frac{r_2}{\alpha[1 - (\frac{1}{10})^s]}$$

We leave it to the student to prove, using the Archimedean law, that such a choice of $\phi(\alpha)$ is possible and conclude that $[a_k - r]$ is a null sequence and hence that $[a_k] \sim [r]$.

Exercises 8.6

1. Convert each of the following repeating decimals into a fraction:
(a) 2.31434343 . . . (b) 0.515515515 . . .
(c) 3.142142142 . . . (d) 6.23414141 . . .

2. Prove, using the Archimedean law, that for any $\alpha \in R^+$, there exists $\phi(\alpha) \in N$ such that when $n \geq \phi(\alpha)$,

$$(10^s)^n > \frac{r_2}{\alpha[1 - (\frac{1}{10})^s]}$$

7. $\sqrt{2}$ as a Real Number

Since one of our motivations for introducing real numbers was the nonexistence of a rational number r such that $r^2 = 2$, it is in order here to show that there is indeed an element $U \in R^*$ such that $U^2 = \{[2]\}$. To do this we define the sequence $[u_k]$ as follows:

$$u_1 = 1, \qquad u_2 = \frac{u_1 + (2/u_1)}{2}, \qquad u_3 = \frac{u_2 + (2/u_2)}{2}$$

and, in general,

$$u_{k+1} = \frac{u_k + (2/u_k)}{2} \quad (k \in N)$$

(This type of definition is known as a *recursive* definition. It yields $u_1 = 1$, $u_2 = 1.5$, $u_3 = 17/12$ etc.) We shall prove that $[u_k]$ is a Cauchy sequence such that if $U = \{[x_k] | [x_k] \sim [u_k]\} \in R^*$, then $U \times U = U^2 = \{[2]\}$ $(\leftrightarrow 2 \in R)$.

We proceed to the proof of $U^2 = \{[2]\}$ in several stages, the details of which we leave to the student.

(1) For all $k \in N$, $u_k > 0$ and $u_k^2 \neq 2$ since each $u_k \in R$.

(2) For all $k \in N$, $u_{k+1}^2 - 2 = \left(\dfrac{u_k^2+2}{2u_k}\right)^2 - 2 = \left(\dfrac{u_k^2-2}{2u_k}\right)^2.$

(3) By (1) and (2), for all $k \in N$ such that $k \geq 2$, $u_k^2 > 2$.

(4) For all $k \in N$, $u_k \geq 1$ since $u_1 = 1$ and if $u_k < 1$ for any $k > 1$, it would follow that $u_k^2 < 1$ contrary to (3).

(5) For all $k \in N$, $0 < u_{k+1}^2 - 2 \leq 1/2^{k+1}$.

Proof of (5). Now $u_{k+1}^2 - 2 > 0$ by (3) and we proceed to prove the second part of our inequality by induction. First, if $k = 1$ our inequality becomes

$$u_2^2 - 2 \leq \tfrac{1}{2^2} = \tfrac{1}{4}$$

But $u_2^2 - 2 = (3/2)^2 - 2 = 1/4$ so that our inequality holds when $k = 1$. Now assume that the inequality holds for $k = n$ so that

$$u_{n+1}^2 - 2 \leq 1/2^{n+1}$$

Now

$$u_{n+2}^2 - 2 = \left(\frac{u_{n+1}^2 - 2}{2u_{n+1}}\right)^2 \leq \left(\frac{u_{n+1}^2 - 2}{2}\right)^2$$

by (2) and (4). But then, by our induction hypothesis,

$$u_{n+2}^2 - 2 \leq \left(\frac{1}{2}\right)^2 \left(\frac{1}{2^{n+1}}\right)^2 = \left(\frac{1}{2^{n+2}}\right)^2 < \frac{1}{2^{n+2}}$$

Thus our inequality holds for $k = 1$ and, if it holds for $k = n$, it holds for $k = n + 1$. By induction it then follows that it holds for all $k \in N$.

(6) If $\alpha \in R^+$, then there exists a $t \in N$ such that $1/2^{t+1} < \alpha$.

Proof of (6). We take $x = 1$ and $y = 1/\alpha$ in Theorem 6.13 (the Archimedean law) and have that there exists $s \in N$ such that $s \cdot 1 = s > 1/\alpha$. Now let $S = \{n | n \in N$ and there exists $t \in N$ such that $2^{t+1} > n\}$. We wish to show that $S = N$ and thus prove that, no matter how large n is, we can find a $t \in N$ such that $2^{t+1} > n$. Now $1 \in S$ since we may take $t = 1$ and have $2^{1+1} > 1$. Suppose now that $k \in S$. Then there exists $t \in N$ such that $2^{t+1} > k$. But then $2 \cdot 2^{t+1} > 2k$ so that

$$2^{(t+1)+1} > k + k \geq k + 1$$

and since $t + 1 \in N$ we have shown that, if $k \in S$, $k + 1 \in S$. Thus $S = N$.

Since $S = N$, for $s > 1/\alpha$ as chosen above, we may find a $t \in N$ such that

$$2^{t+1} > s > 1/\alpha$$

Hence $1/2^{t+1} < \alpha$ as desired.

(7) The sequence $[u_k^2] - [2] = [u_k^2 - 2]$ is a null sequence.

Proof of (7). By (5) we have

$$|u_{n+1}^2 - 2| \leq 1/2^{n+1}$$

Now for any $\alpha \in R^+$ we take $\phi(\alpha) = t$ for t as determined in (6) and have, for $n \geq \phi(\alpha) = t$,

$$|u_{n+1}^2 - 2| \leq 1/2^{n+1} < \alpha$$

since, if $n \geq t$, $(1/2^{n+1}) \leq (1/2^{t+1})$.

(8) $[u_k^2]$ is a Cauchy sequence.

Proof of (8). By Theorem 5, $[u_k^2 - 2]$ is a Cauchy sequence since it is a null sequence. Hence for any $\alpha \in R^+$ there exists $\phi(\alpha) \in N$ such that when $n, m \geq \phi(\alpha)$,

$$|(u_n^2 - 2) - (u_m^2 - 2)| = |u_n^2 - u_m^2| < \alpha$$

Thus $[u_k^2]$ is a Cauchy sequence.

(9) $[u_k]$ is a Cauchy sequence.

Proof of (9). $|u_n^2 - u_m^2| = |u_n - u_m||u_n + u_m| = |u_n - u_m|(u_n + u_m)$ since $u_k > 0$ for all $k \in N$. Now since $[u_k^2]$ is a Cauchy sequence, for any $\alpha \in R^+$ we may obtain $\phi(2\alpha) \in N$ such that for $n, m \geq \phi(2\alpha)$, $|u_n^2 - u_m^2| < 2\alpha$. Hence $|u_n - u_m|(u_n + u_m) < 2\alpha$ so that for $n, m \geq \phi(2\alpha)$,

$$|u_n - u_m| < \frac{2\alpha}{u_n + u_m} \leq \frac{2\alpha}{2} = \alpha$$

since $u_k \geq 1$ for all $k \in N$ by (4). Thus $[u_k]$ is a Cauchy sequence.

(10) We now consider $\{[u_k]\}$ and have $\{[u_k]\}^2 = \{[u_k^2]\} = \{[2]\}$ by (7) and our definition of equality for real numbers.

Exercises 8.7

1. Use the sequence $[u_k]$ defined by

$$u_1 = 1, \; u_{k+1} = \frac{u_k + (3/u_k)}{2}$$

to prove that R^* contains $\sqrt{3}$.

2. Generalize the method used in establishing the existence of $\sqrt{2}$ and $\sqrt{3} \in R^*$ to establish the existence of $\sqrt{a}$ for any $a \in R^+$.

3. Use the sequence $u_1 = 1$, $u_{k+1} = [1/(u_k + 1)] + 1$ to prove the existence of $\sqrt{2} \in R^*$.

8. Ordering of the Real Numbers

Our intuitive feeling of a positive sequence is one in which the terms "ultimately" become and remain positive. Thus, for example, our feeling is that we should consider the sequence

$$-4, -2, 1, 1\tfrac{1}{2}, 1\tfrac{1}{4}, \ldots, 1+1/2^{n-3}, \ldots$$

positive even though some of the terms are negative. We formalize this feeling in our next definition.

Definition 11. *A Cauchy sequence $[a_k]$ is said to be* **positive** *if and only if there exists $r \in R^+$ and $n \in N$ such that for all $m \in N$, $m \geq n$ implies $a_m \geq r$. We write $[a_k] > [0]$ to mean that $[a_k]$ is positive.*

Thus in our example above we may take $r = 1$ and $n = 3$.

For convenience we will call the ordered pair (n, r) in Definition 11 a *test pair* for the sequence $[a_k]$. We may thus say that a Cauchy sequence is positive if and only if it has a test pair.

Theorem 11. If $[a_k]$ and $[b_k]$ are positive Cauchy sequences, then so are $[a_k] + [b_k]$ and $[a_k][b_k]$.

Proof. Let (n_1, r_1) be a test pair for $[a_k]$ and (n_2, r_2) be a test pair for $[b_k]$. Then for all $m \geq \max(n_1, n_2) = n$, $a_m + b_m \geq r_1 + r_2 = r$. Hence (n, r) is a test pair for $[a_k] + [b_k]$ and $[a_k] + [b_k]$ is a positive Cauchy sequence.

Similarly, the student may check that $(n, r_1 r_2)$ is a test pair for $[a_k][b_k]$ and hence conclude that $[a_k][b_k]$ is also a positive Cauchy sequence.

Definition 12. *A Cauchy sequence $[a_k]$ is said to be* **negative** *if and only if $-[a_k] = [-a_k]$ is positive. We write $[a_k] < [0]$ to mean that $[a_k]$ is negative.*

Theorem 12. (Trichotomy law for Cauchy sequences) If $[a_k]$ is a Cauchy sequence, then one and only one of the following hold:

$$(1)\ [a_k] > [0] \qquad (2)\ [a_k] \sim [0] \qquad (3)\ [a_k] < [0]$$

Proof. We proceed by showing in turn that
(a) If (2) holds, neither (1) nor (3) can hold;
(b) (1) and (3) cannot both hold;
(c) If neither (1) nor (3) holds, then (2) holds.
From (a) and (b) we will then conclude that at *most* one of (1), (2), and (3)

holds, and from (c) that at *least* one of (1), (2), and (3) holds. Thus the establishment of (a), (b), and (c) will provide a proof of the theorem.

We now turn to the proofs of (a), (b), and (c).

(a) Suppose $[a_k] \sim [0]$ and also $[a_k] > [0]$. We then let (m, r) be a test pair for $[a_k]$. Since $[a_k] \sim [0]$, there exists $\phi(r/2) \in N$ such that when $n \geq \phi(r/2)$, $a_n < r/2 \in R^+$. Now choose $n \geq max\ \{m, \phi(r/2)\}$ and have $a_n < r/2$ because $[a_k] \sim [0]$ and also $a_n \geq r$ because $[a_k] > [0]$. From this contradiction we conclude that if $[a_k] \sim [0]$, then we cannot have $[a_k] > [0]$.

On the other hand, if $[a_k] \sim [0]$ and also $[a_k] < [0]$ we observe that then $[-a_k] \sim [0]$ and $[-a_k] > [0]$ so that we again have a contradiction.

(b) If $[a_k] > [0]$ and also $[a_k] < [0]$ it follows that $[a_k] > [0]$ and also $[-a_k] > [0]$. But then, by Theorem 11, $[a_k] + [-a_k] > [0]$ and since $[a_k] + [-a_k] = [0]$, we have $[0] > [0]$, contrary to (a).

(c) Suppose $[a_k] \ngtr [0]$ and $[a_k] \nless [0]$. Then $[-a_k] \ngtr [0]$ for if $[-a_k] > [0]$ it would follow by Definition 12 that $[-(-a_k)] = [a_k] < [0]$, a contradiction. Since $[a_k]$ is a Cauchy sequence we know that for any $r \in R^+$ there exists $\phi(r/2) \in N$ such that when $n, m \geq \phi(r/2)$, we have $|a_n - a_m| < r/2$. But since $[a_k] \ngtr [0]$ it has no test pair and, in particular, $(\phi(r/2), r/2)$ is not a test pair. Hence there exists a natural number $m \geq \phi(r/2)$ such that $a_m < r/2$. We now write

$$a_n = a_m + (a_n - a_m) \leq a_m + |a_n - a_m|$$

and have, for $n \geq \phi(r/2)$

$$a_n < (r/2) + (r/2) = r$$

for any $r \in R^+$. But, by the same argument, we may use the fact that $[-a_k] \ngtr [0]$ to obtain $-a_n < r$ and hence that, for $n \geq \phi(r/2)$, $|a_n| = |a_n - 0| < r$. Thus $[a_n] \sim [0]$.

Theorem 13. If $[a_k] > [0]$ and $[b_k] \sim [0]$, then $[a_k] + [b_k] > [0]$. Similarly if $[a_k] < [0]$ and $[b_k] \sim [0]$, then $[a_k] + [b_k] < [0]$.

Proof. Suppose $[a_k] > [0]$ and consider $[c_k] = [a_k] + [b_k]$. By the trichotomy law (Theorem 12), $[c_k] > [0]$, $[c_k] \sim [0]$, or $[c_k] < [0]$. But if $[c_k] \sim [0]$, then (Theorem 6), $[a_k] = [c_k] - [b_k] \sim [0]$ which is a contradiction to the trichotomy law. Similarly, if $[c_k] < [0]$ we have $[-c_k] > [0]$ and hence, by Theorem 11, $[-c_k] + [a_k] > [0]$. But then $[b_k] = -([-c_k] + [a_k]) < [0]$ which is again a contradiction to the trichotomy law. Hence $[c_k] > [0]$.

The proof of the second part of the theorem and the following corollary are left as exercises for the student.

Corollary. If $[a_k] > [0]$ and $[b_k] \sim [a_k]$, then $[b_k] > [0]$. Similarly, if $[a_k] < [0]$ and $[b_k] \sim [a_k]$, then $[b_k] < [0]$.

We are now in a position to define the concept of positiveness for real numbers.

Definition 13. *Let* $A = \{[x_k] | [x_k] \sim [a_k]\} = \{[a_k]\} \in R^*$. *Then* A *is said to be* **positive** *and we write* $A > \bar{0}$ *if and only if* $[a_k] > [0]$.

Suppose $A = \{[a_k]\} = \{[b_k]\}$. Then $[a_k] \sim [b_k]$ so that, by the corollary to Theorem 13, $[a_k] > [0]$ if and only if $[b_k] > [0]$. That is, the question of the positiveness of any real number is independent of the sequence used to represent the equivalence class that is the real number; hence we may pass freely from a real number to any equivalence class representative in considering questions concerning positiveness of real numbers.

Theorem 14. If $A, B \in R^*$ and $A, B > \bar{0}$, then $A + B > \bar{0}$ and $AB > 0$.

Proof. Let $A = \{[a_k]\}$ and $B = \{[b_k]\}$ where $[a_k] > [0]$ and $[b_k] > [0]$. Then $A + B = \{[a_k] + [b_k]\}$ and $AB = \{[a_k][b_k]\}$ where $[a_k] + [b_k] > [0]$ and $[a_k][b_k] > [0]$ by Theorem 11. Hence $A + B > \bar{0}$ and $AB > \bar{0}$.

We leave the trichotomy law for real numbers as an exercise for the student to prove using Theorem 12.

Theorem 15. If $A \in R^*$, then one and only one of the following alternatives hold: (1) $A > \bar{0}$, (2) $A = \bar{0}$, or (3) $-A > \bar{0}$.

Definition 14. *Let* $A - B = A + (-B)$ *for* $A, B \in R^*$, *and define* $A > B$ *if and only if* $A - B > \bar{0}$ *and* $A < B$ *if and only if* $B - A > \bar{0}$.

Theorem 16. If $A, B \in R^*$ one and only one of $A > B$, $A = B$, $B > A$ hold.

Proof. This follows directly from Theorem 15 and Definition 14.

We leave as a series of exercises for the student the proofs of various properties of inequalities for real numbers such as if $A, B, C \in R^*$ and $A > B$, then $A + C > B + C$. In each case the proofs follow the corresponding proofs of inequalities involving rational numbers.

Exercises 8.8

1. Complete the proof of Theorem 11.
2. Complete the proof of Theorem 13.
3. Prove the corollary to Theorem 13.
4. Prove Theorem 15.

5. If A, B, C, and D are real numbers, prove
   (a) If $A > B$, then $A + C > B + C$;
   (b) If $A > B$, and $C > D$, then $A + C > B + D$;
   (c) If $A > B$, and $C > \bar{0}$, then $AC > BC$.

9. The Field of Real Numbers

The student will recall (Theorem 9) that we have shown that R^* is a commutative ring with a multiplicative identity $\bar{1} = \{[1]\}$. Thus to show that R^* is a field we need only to show that every nonzero element of R^* has a multiplicative inverse. (Cf. Definitions 5.4 and 7.3.)

The basic result we need to establish this result is embodied in the following lemma.

Lemma. Suppose $[a_k]$ is a Cauchy sequence and $[a_k] > [0]$ with test pair (t, r). Then if $[b_k]$ is defined by $b_k = 0$ for $k < t$ and $b_k = a_k^{-1}$ if $k \geq t$, we have (1) $[b_k]$ is a Cauchy sequence and (2) $[a_k][b_k] \in \{[1]\}$.

Proof. (1) Since $[a_k]$ is a Cauchy sequence, for any $s \in R^+$ we know that there exists $\phi(sr^2) \in N$ such that if $n, m \in N$ and $n, m \geq \phi(sr^2)$ we have $|a_n - a_m| < sr^2$. Now let $\sigma(s) = max\,\{t, \phi(sr^2)\}$. Then if $n, m \geq \sigma(s)$ we have

$$|b_n - b_m| = |a_n^{-1} - a_m^{-1}| = \left|\frac{a_m - a_n}{a_n a_m}\right| = \frac{|a_m - a_n|}{a_n a_m} < \frac{sr^2}{r^2} = s$$

Hence $[b_k]$ is a Cauchy sequence.

(2) Consider the sequence $[a_k][b_k] - [1]$. Now $a_k b_k - 1 = 0$ for all $k \geq t$ and hence it is obvious that $[a_k][b_k] - [1] \sim [0]$ and thus that $[a_k][b_k] \in \{[1]\}$.

Theorem 17. R^* is a field.

Proof. Suppose $A = \{[a_k]\} \in R^*$ and $A \neq 0$. Then $[a_k] \nsim [0]$. If $[a_k] > [0]$ we invoke our lemma to conclude that there exists a Cauchy sequence $[b_k]$ such that $[a_k][b_k] \sim [1]$. Now we let $B = \{[b_k]\}$ and have $AB = \{[a_k][b_k]\} = \{[1]\}$ as desired.

If $[a_k] < [0]$, then $[-a_k] > [0]$ and we again invoke our lemma to obtain a Cauchy sequence $[-b_k]$ such that $[-a_k][-b_k] \sim [1]$. Finally, since $[-a_k][-b_k] = [a_k][b_k]$ we have $AB = \{[a_k][b_k]\} = \{[1]\}$ as before.

Example. Let $A = \{[a_k]\}$ where $a_1 = -1000$, $a_2 = 0$, $a_3 = 17/5$, $a_4 = 1$, and

$$a_{k+1} = \frac{a_k + 2/a_k}{2} = \frac{a_k^2 + 2}{2a_k}$$

for $k \geq 4$. Then if $b_1 = b_2 = 0$, $b_3 = 5/17$, $b_4 = 1$,

$$b_{k+1} = \frac{2a_k}{a_k^2 + 2} \text{ for } k \geq 4,$$

and $B = \{[b_k]\}$, we have $AB = \{[a_k][b_k]\} = \{[1]\}$.

Exercises 8.9

1. Find the multiplicative inverses of the following real numbers $\{[a_k]\}$ where
   (a) $a_1 = 2/3$, $a_2 = 1/5$, $a_3 = -2$, $a_k = (2k + 1)/3k$ for $k > 3$
   (b) $a_1 = 2$, $a_{k+1} = (a_k^2 + 3)/3a_k$ for $k \geq 1$
   (c) $a_1 = 2$, $a_2 = 0$, $a_3 = -7/3$, $a_k = -(3k + 1)/7k$ for $k > 3$.
2. Prove that if $A \in R^*$, then $A^2 \geq \bar{0}$.

10. Additional Remarks

(1) The same process of forming equivalence classes of Cauchy sequences may be used with any ordered field in place of R. The field obtained is called the *completion* of the given field. Thus R^* is the completion of R. (See, for example, [B4].)

(2) Suppose we begin anew and define in a similar fashion Cauchy sequences $[A_k]$ with $A_k \in R^*$. We can then form equivalence classes of these sequences, etc., and obtain a field $\bar{R}$ that is the completion of R^*. It may be proved, however, that $\bar{R}$ is isomorphic to R^*. Any ordered field whose completion is isomorphic to itself is called a *complete* field. Thus R^* is a complete field but R is not. (See, for example, [H3].)

(3) At the beginning of this chapter we remarked that it was *not* possible to construct the real numbers as equivalence classes of pairs of rational numbers. Let us conclude our work on the real numbers by indicating briefly why this is so.

First we need the concept of equinumerous sets.

Definition 15. *Two sets are said to be* ***equinumerous*** *if and only if there exists a* 1–1 *correspondence between the elements of the two sets.*

For finite sets this is a simple concept and corresponds to the ordinary idea of "the same numbers of elements". Thus there are three elements in the set $\{a, b, c\}$ and three elements in the set $\{A, B, C\}$ so that the two sets are equinumerous as is indicated by the correspondence $a \leftrightarrow A$, $b \leftrightarrow B$, and $c \leftrightarrow C$. On the other hand, the two sets $\{a, b, c\}$ and $\{A, B, C, D\}$ are not equinumerous.

The situation is not nearly as simple for infinite sets since, by definition, we cannot "count" the number of elements as we can do for finite sets. Off hand, however, we might be inclined to think that the two infinite sets

$$N = \{n|n \text{ a natural number}\} = \{1, 2, 3, \ldots\}$$

and

$$M = \{2n|n \text{ a natural number}\} = \{2, 4, 6, \ldots\}$$

were not equinumerous since $M \subseteq N$ and $M \neq N$. Nevertheless, according to our definition, M and N are equinumerous as is shown by the 1–1 correspondence defined by

$$n \in N \leftrightarrow 2n \in M$$

which associates, in a definite fashion, a unique element of M with any element of N and, conversely, a unique element of N with every element of M. That is, there is a 1–1 mapping, σ, of N onto M where $n\sigma = 2n \in M$ for all $n \in N$.

Furthermore the set of I of integers is also equinumerous with N as is shown by the 1–1 correspondence defined as follows. Let n be any element of N and let $n \in I \leftrightarrow 2n - 1 \in N$, $0 \in I \leftrightarrow 2 \in N$, $-n \in I \leftrightarrow 2n + 2 \in N$. Thus $1 \in I \leftrightarrow 1 \in N$, $2 \in I \leftrightarrow 3 \in N$, $3 \in I \leftrightarrow 5 \in N, \ldots, 0 \in I \leftrightarrow 2 \in N$, $-1 \in I \leftrightarrow 4 \in N$, $-2 \in N \leftrightarrow 6 \in N, \ldots$.

We next show that, even more surprisingly, the set, R^+, of positive rational numbers is equinumerous with N. To do this imagine that all of the fractions representing positive rational numbers are arranged as indicated in the following diagram (ignore the dotted line for the moment):

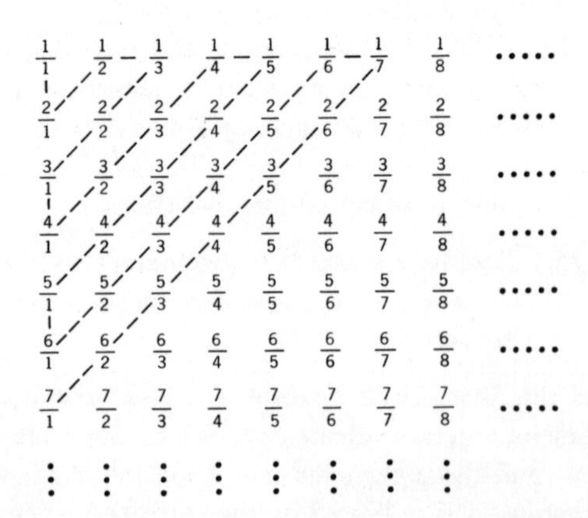

We note, of course, that we have repetition since $1/1 = 2/2 = 3/3 = \ldots$; $2/3 = 4/6 = 6/9 = \ldots$, etc.

Now suppose that we have given $n \in N$ and wish to describe a rule for associating with this $n \in N$ a definite rational number $r \in R^+$. What we do is to follow the dotted line in the diagram that begins at 1/1, goes down one line to 2/1, then diagonally up until it reaches the edge, then to the right one line, down diagonally until it reaches the edge, down one line, up diagonally, etc. As we follow the line we count $1, 2, 3, \ldots, n$ as we go through each fraction except that we never count repetitions. For example, suppose that we have $n = 5$. We have 1/1, 2/1, 1/2, 1/3, 3/1 (omitting $2/2 = 1/1$) and hence have $5 \in N \to 3/1 \in R^+$. Similarly, if $n = 13$ we obtain 1/1, 2/1, 1/2, 1/3, 3/1, 4/1, 3/2, 2/3, 1/4, 1/5, 5/1, 6/1, 5/2 (omitting $2/2 = 1/1$, $2/4 = 1/2$, $3/3 = 1/1$, and $4/2 = 2/1$). Thus $13 \in N \to 5/2 \in R^+$.

Conversely, given any $r \in R^+$ we take its representation as a fraction in lowest terms and count "how far" it is from 1/1 along the dotted line (again omitting repetitions). Thus if we have given $4/6 \in R^+$ we observe that $4/6 = 2/3$ and count 1/1, 2/1, 1/2, 1/3, 3/1, 4/1, 3/2, 2/3 so that we have $2/3 \in R^+ \to 8 \in N$. Clearly our correspondence is 1–1 and thus R^+ and N are equinumerous.

Next we observe that the set of ordered pairs of positive rational numbers is equinumerous with N. Our procedure in establishing this fact is similar to that used to show that R^+ is equinumerous with N. Thus since R^+ is equinumerous with N we can write $r_1 = 1/1$, $r_2 = 2/1$, $r_3 = 1/2$, $r_4 = 1/3$, $r_5 = 3/1$, $r_6 = 4/1$, $r_7 = 3/2$, and so on as dictated by the correspondence just described between R^+ and N. Now we arrange the ordered pairs of elements of R^+ as

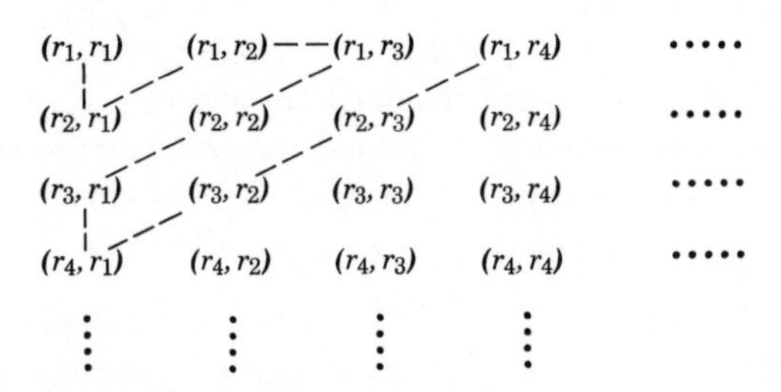

and proceed as before along the dotted line to determine the matching. Thus, for example, $3 \in N \to (r_1, r_2) \in R^+ \times R^+$.

Because of the fact that I is equinumerous with N it is certainly plausible that R is equinumerous with R^+ and that the set of ordered pairs of elements of R is equinumerous with the set of ordered pairs of elements of R^+; we leave the proofs to the student. Hence if it were possible to represent the real numbers as pairs of rational numbers, it would follow that R^* would be equinumerous with N. We will show that this is not the case and, indeed, that even the set of real numbers r with $0 \leq r \leq 1$ is not equinumerous with N.

We recall that a real number, r, is a set of equivalent Cauchy sequences: $r = \{[x_k] | [x_k] \sim [a_k]\}$. For our purposes here, however, it will do no harm to regard a real number as a single sequence as long as we take care not to list two equivalent sequences. In particular, let us consider the sequence $[r_k]$ where $r_1 = 0.a_1, r_2 = 0.a_1a_2, \ldots, r_k = 0.a_1a_2 \ldots a_k$ where $a_1, a_2, a_3, \ldots$ are in the set $\{n | n \in N \text{ and } 0 \leq n \leq 9\}$.

We leave it to the student to prove that such sequences are Cauchy sequences and hence are representatives of the class of sequences that are our real numbers. We also leave to the student to prove that the only such sequences that are equivalent (but not equal) are those in which we have eventually a repetition of 9's on the one hand and a repetition of 0's on the other; e.g., 0.32000 . . . and 0.31999 . . . . To avoid such situations we will agree not to use as our representatives such sequences that end in a repetition of 9's.

Now let P be the set of all such sequences $[r_k]$ as just described and suppose that we do have a 1–1 correspondence of P with N as indicated below:

$$1 \in N \leftrightarrow 0.a_1a_2a_3 \ldots = a \in P$$
$$2 \in N \leftrightarrow 0.b_1b_2b_3 \ldots = b \in P$$
$$3 \in N \leftrightarrow 0.c_1c_2c_3 \ldots = c \in P$$
$$4 \in N \leftrightarrow 0.d_1d_2d_3 \ldots = d \in P$$
$$5 \in N \leftrightarrow 0.e_1e_2e_3 \ldots = e \in P$$

Now consider a number $s \in P$ where $s = 0.a_1'b_2'c_3'd_4' \ldots$ such that $a_1' \neq a_1$, $b_2' \neq b_2$, $c_3' \neq c_3$, $d_4' \neq d_4$, ... Since $a_1' \neq a_1$, $s \neq a$ and thus $s \not\leftrightarrow 1$; since $b_2' \neq b_2$, $s \neq b$ and thus $s \not\leftrightarrow 2$; since $c_3' \neq c_3$, $s \neq c$ and thus $s \not\leftrightarrow 3$; . . . ; $s \not\leftrightarrow n$ for any $n \in N$. Hence our given correspondence is not 1–1 and since it was a completely arbitrary correspondence we conclude that no such 1–1 correspondence exists between P and N.

Exercises 8.10

1. Consider the correspondence given between N and R^+. Determine the numbers in R^+ corresponding to the following numbers in N.
(a) 6 (b) 10
(c) 17 (d) 21.

2. Consider the correspondence given between N and R^+. Determine the numbers in N corresponding to the following numbers in R^+.
(a) 3/4 (b) 5/8
(c) 12/14 (d) 10/15.

3. Let A, B, and C be sets. If A and B are equinumerous and B and C are equinumerous, prove that A and C are equinumerous.
4. Prove that R is equinumerous with R^+.
5. Let A be a set equinumerous with N and $B = \{(a, b) | a, b \in A\}$. Prove that A and B are equinumerous.
6. Suppose that the correspondence of natural numbers to positive real numbers as given on page 122 is restricted to rational numbers; i.e., $0.a_1a_2a_3, \cdots, 0.b_2b_2b_2 \cdots$ etc. are rational numbers. Why can we not conclude by a similar argument that the rational numbers are not equinumerous with N?

*7. Prove that a sequence $[r_k]$ where $r_k = 0.a_1a_2 \ldots a_k$ and $a_i \in \{n | n \in N$ and $0 \leq n \leq 9\}$ for all $i \in N$ is a Cauchy sequence.

*8. Prove that if r is a real number such that $\bar{0} \leq r < 1$, then $r = \{[r_k]\}$ where r_k is as described in problem 7.

*9. Prove that two sequences $[r_k]$ and $[r_k']$ of the type described in problem 7 are equivalent if and only if either $a_i = a_i'$ for all $i \in N$ or there exists $n \in N$ such that, for $1 \leq i < n$, $a_i = a_i'$, $a_n' = a_n - 1$ and for $i > n$, $a_i = 0$ and $a_i' = 9$.

*10. A set is said to be a *countable* or *denumerable* set if it is equinumerous with N.
   (a) If S and T are two countable sets, prove that $S \cup T$ is countable.
   (b) If $S = T \cup V$ where S is countable and T is finite, prove that V is countable.
   (c) Prove that the ring $R[x]$ of all polynomials with integral coefficients is countable.

PROBLEMS FOR FURTHER INVESTIGATION

1. Our development of the real numbers depends strongly on the concept of absolute value: $|a| = a$ if $a \geq 0$, $|a| = -a$ if $a < 0$. Consider now any positive prime number p. Then any rational number r can be written as $r = p^\alpha(a/b)$ where α, a, and b are integers such that $(ab, p) = 1$. (For example, if $p = 3$, $9/10 = 3^2(1/10)$, $4/15 = 3^{-1}(4/5)$, $5/7 = 3^0(5/7)$.) Define $|r|_p = (1/2)^\alpha$. (For example, $|9/10|_3 = (1/2)^2 = 1/4$, $|4/15|_3 = (1/2)^{-1} = 2$, $|5/7|_3 = (1/2)^0 = 1$.) Does this (*p-adic*) absolute value have the properties of ordinary absolute value? Can you use it to define (p-adic) null sequences, etc.? What will p-adic numbers look like if they are derived from R by using $|r|_p$ as R^* is derived from R by using $|r|$? [H5], [J3], [V1]
2. A *Dedekind cut*, C, in R is a division of R into two classes $A \neq \varnothing$ and $B \neq \varnothing$ such that $A \cup B = R$ and if $x \in A$ and $y \in B$, then $x < y$. We write $C = (A|B)$. For example we might have $A = \{x \in R | x \leq 2\}$ and $B = \{x \in R | x > 2\}$ or $A = \{x \in R | x^2 \leq 2\}$ and $B = \{x \in R | x^2 > 2\}$. In some sense, then, the first cut corresponds to the (rational) number, 2, and the second cut, corresponds to the (irrational) number, $\sqrt{2}$. Can you frame an appropriate definition of a real number in terms of cuts and develop an arithmetic of these real numbers? [B4], [E2]
3. A set S will be said to have (finite) *cardinal number* $n \in N$ if and only if there exists a 1–1 correspondence between the elements of S and the set $\{1, 2,$

. . . , $n\}$. There is, of course, a well-established arithmetic of finite cardinal numbers (commonly studied in the elementary school). Can you define infinite cardinal numbers and establish an arithmetic for these? [B4]

REFERENCES FOR FURTHER READING

[B4], [E2], [H3], [K2], [L1], [R1].

9

Rings, Ideals, and Homomorphisms

We have already defined a ring and discussed some of the elementary properties of rings in Section 5.2. For convenience, however, let us repeat here the definition of a ring (in slightly altered form).

Definition 1. *A **ring** is a set S on which are defined two binary operations, $+$ and $\times$, such that S is an Abelian group under $+$ and such that $a, b, c \in S$ implies*

$$a \times (b + c) = (a \times b) + (a \times c),$$

$$(b + c) \times a = (b \times a) + (c \times a),$$

$$a \times (b \times c) = (a \times b) \times c$$

As usual we write ab for $a \times b$.

Our previous examples of rings included integral domains and fields, the set of even integers, the set consisting of 0 alone, and J_4. In Chapter 12 we will study polynomial rings in considerable detail. Here we will be concerned mainly with the concept of an ideal of a ring and the construction of residue class rings and we will confine ourselves to commutative rings—although a considerable part of our results will remain valid for general rings. Henceforth, then, in this chapter the word "ring" is to mean "commutative ring".

1. Subrings and Ideals

Definition 2. *A **subring** R of a ring S is a subset of S which is itself a ring under the operations of S.*

Examples. The set of rational numbers is a subring of the ring of real numbers; the set of integers is a subring of the ring of rational numbers; the set of all polynomials of the form $a_0x^{2n} + a_1x^{2n-2} + \cdots + a_n$ (a_i in a field F) is a subring of the ring $F[x]$ of all polynomials in x with coefficients in F.

Definition 3. *A subring R of a ring S is an **ideal** of S if $s \in S$ and $r \in R$ implies sr $(= rs)$*† *$\in R$. We write SR $(= RS) \subseteq R$.*

Example 1. The subring of even integers is an ideal of the ring of integers since, if n is even, mn is even for any integer m.

Example 2. Let $S = F[x, y]$ be the ring of all polynomials in two indeterminants x and y with coefficients in some field F and let R be the subring of S consisting of all polynomials of the form $xf(x, y)$ where $f(x, y) \in S$. Then R is an ideal of $F[x, y]$.

Example 3. The set R of all polynomials of the form $a_0x^{2n} + a_1x^{2n-2} + \cdots + a_n$ (a_i in a field F) is a subring of the ring $S = F[x]$ but is not an ideal of S since, for example, $x \cdot x^2 = x^3 \notin R$.

Clearly any ring S has at least two ideals: itself and the ring $\{0\}$ consisting only of the additive identity of S. An ideal of a ring S that is not equal to S or $\{0\}$ is called a *proper ideal* of S.

Theorem 1. A ring S that is also a field has no proper ideals.

Proof. Suppose that R is an ideal of S and $R \neq \{0\}$. Then there exists an element $x \in R$ with $x \neq 0$. Since S is a field, x has a multiplicative inverse, x^{-1}, and since R is an ideal of S we have $x \cdot x^{-1} = e \in R$ where e is the multiplicative identity of S. But then $ae = a \in R$ for all $a \in S$ and hence $R = S$.

The ideals of Examples 1 and 2 are both examples of a special class of ideals known as principal ideals. Before we give a formal definition of a principal ideal, however, let us look at a matter of notation. In a ring S we have the notion of a product, xy, for $x, y \in S$. On the other hand, we write $x + x = 2x$, $x + x + x = 3x$ and, in general, the sum of n x's as nx. Now nx looks like a product but is not (unless, of course, the integer n is in S). Rather, nx is simply an abbreviation for the sum of n x's. Similarly, by $-nx$ for $n > 0$ we mean $n(-x)$, the sum of n $(-x)$'s, and by $0x$ where 0 is the integer 0 we mean 0, the additive identity of S.‡

We leave as an exercise for the student to prove that if $m, n \in I$ and $r, s \in S$, then $(nm)s = n(ms)$ and $n(rs) = (nr)s$.

Theorem 2. Let S be a ring and $r \in S$. Then $R_r = \{nr + sr | n \in I, s \in S\}$ is an ideal of S called a *principal ideal*. We write $R_r = [r]$.

Before proving Theorem 2 let us prove a useful lemma concerning ideals.

† This, of course, is for commutative rings. In a noncommutative ring we have left ideals ($sr \in R$) and right ideals ($rs \in R$) and an ideal is a subring that is both a left ideal and a right ideal.

‡ It is customary, even though sometimes confusing, to use "0" as the symbol for the additive identity of any ring.

Lemma 1. Let R be a nonempty subset of S such that
(1) If $x, y \in R$, then $x - y \in R$;
(2) $SR \subseteq R$ where $SR = \{sr | s \in S, r \in R\}$.
Then R is an ideal of S.

Proof. Consider an element x of R. Then $x - x = 0 \in R$ and since $0 - x = -x$ it follows that $-x \in R$ if $x \in R$. Finally, $x + y = x - (-y)$ so that $x, y \in R$ implies $x + y \in R$. Using (2) we have $RR \subseteq SR \subseteq R$. Thus R is a subring of S and, by (2), is an ideal of S.

Now let us prove Theorem 2. Suppose $x, y \in R_r$. Then $x = n_1 r + s_1 r$, $y = n_2 r + s_2 r$, and $x - y = (n_1 - n_2)r + (s_1 - s_2)r \in R$. Also, if $s \in S$ we have

$$sx = (n_1 r)s + (ss_1)r = (n_1 s + ss_1)r = 0r + (n_1 s + ss_1)r \in R$$

since $n_1 s + ss_1 \in S$. Hence R_r is an ideal of S by the lemma.

If S is a ring with an identity, e, for multiplication we may write $nr + sr$ as $n(er) + sr = (ne)r + sr = (ne + s)r$ so that a principal ideal, $[r]$, in such a ring may be written in the simpler form $\{sr | s \in S\}$ and we note, in particular, that $[e] = S$. Thus in Example 1 the ideal consisting of the even integers is equal to $[2]$ and, in Example 2, the ideal R is equal to $[x]$. But if S does not have a multiplicative identity the two forms are not equivalent. For example, if S is the ring of even integers and R the ideal consisting of integers of the form $4k$ $(k \in I)$ we have $R = [4]$. In particular, $4 \in R$ since $4 = 1(4) + 0 \cdot 4$ (where 1 is understood as indicating one 4 rather than as 1 multiplied by 4). But $4 \neq s \cdot 4$ for some even integer s, i.e., $4 \notin \{s \cdot 4 | s \in S\}$.

Not all ideals are principal ideals. To see this consider the ring $S = F[x, y]$ where F is any field. (See Example 2, p. 126.) Since S has a multiplicative identity (which is the multiplicative identity of F) we know that every principal ideal of S is of the form $R_r = \{sr | s \in S\}$. Now let R be the set of all polynomials with zero constant terms; i.e., $f(x, y) \in R$ if and only if $f(0, 0) = 0$. It is easy to see that R is an ideal of S, and we now claim that it is *not* a principal ideal.

To prove this, suppose that it is a principal ideal; that is, suppose that $R = [f(x, y)]$. We note first that $f(x, y)$ is not of zero degree since a polynomial of zero degree is an element, d, of F. But then either $d = 0$ in which case $R = [0]$, a contradiction, or d has an inverse, d^{-1}, and $d^{-1}d = e$, the identity element of F. In this case, however, $R = S$, again a contradiction.

Next we note that $x \in R$ and $y \in R$ so that if $R = [f(x, y)]$ there must exist $g(x, y) \in S$ and $h(x, y) \in S$ such that

$$(1) \quad x = g(x, y)f(x, y) \qquad \text{and} \qquad y = h(x, y)f(x, y)$$

Now the degree of the product of two nonzero polynomials is clearly the sum of the degrees of the factors. (Formal proof in Theorem 2 of Chapter

12!) Hence (1) implies that $g(x, y)$ and $h(x, y)$ are of degree 0 and that $f(x, y)$ is of the first degree. That is, $f(x, y) = a_1x + a_2y$, $g(x, y) = b$, and $h(x, y) = c$ for $a_1, a_2, b, c \in F$. Thus

$$x = b(a_1x + a_2y) \qquad \text{and} \qquad y = c(a_1x + a_2y)$$

so that $ba_1 = 1$, $ba_2 = 0$, $ca_1 = 0$, and $ca_2 = 1$. Since $b \neq 0$, $a_2 = 0$ and therefore $ca_2 \neq 1$. We conclude that R cannot be a principal ideal.

Exercises 9.1

1. Prove that if S is a ring, $r, s \in S$, and $n, m \in I$, then $(nm)s = n(ms)$ and $n(rs) = (nr)s$. (*Hint*: For the first part consider S as a group under addition.)
2. Prove that the set of integral multiples of any fixed integer m in the ring I is an ideal. For what integers, m, do we have a principal ideal?
3. Find all the ideals (a) of J_6; (b) of J_{10}.
4. In the ring $I[x]$ define $R = \{f(x) | f(x) \in I[x], \text{ and } f(\sqrt{2}) = 0\}$. Prove that $R = [x^2 - 2]$.
5. If S is an integral domain and $r, s \in S$, prove that $[r] = [s]$ if and only if r and s are associates (Definition 5.9).
6. Let $S = F[x, y] = \{a + b_1x + b_2y + c_1x^2 + c_2xy + c_3y^2 + \cdots | a, b_i, c_i, \cdots \in F\}$ where F is any field. Which of the following sets of polynomials are ideals of S?
   (a) All $f(x, y)$ with nonzero constant term ($a \neq 0$);
   (b) All $f(x, y)$ not involving x ($b_1 = c_1 = c_2 = \cdots = 0$);
   (c) All $f(x, y)$ with no constant or linear term ($a = b_1 = b_2 = 0$);
   (d) All $f(x, y)$ without a quadratic term (x^2, y^2, or xy) ($c_1 = c_2 = c_3 = 0$).
7. We may generalize the concept of a principal ideal, $[r]$, of a ring S to the ideal generated by $r_1, r_2, \ldots, r_n$ where $r_1, r_2, \ldots, r_n \in S$. We write

   $$[r_1, r_2, \ldots r_n] = \left\{ \sum_{j=1}^{n} n_j r_j + \sum_{j=1}^{n} s_j r_j \,\middle|\, n_j \in I, s_j \in S \right\}$$

   and call $\{r_1, r_2, \ldots, r_n\}$ a *basis* of the ideal.
   (a) Show that when $n = 1$ we obtain a principal ideal;
   (b) Prove that $[r_1, r_2, \ldots, r_n]$ is an ideal of S;
   (c) If $S = I$, prove that $[6, 4] = [2]$ and that $[9, 25] = S$. Generalize these examples.
   (d) In $R[x]$, prove that $[x^2 + 5x + 6, x + 3] = [x + 3]$ and $[x^2 + 1, x + 3] = R[x]$. Generalize these examples.
8. If D is an integral domain and a ring R is a subset of D, is R necessarily an integral domain?

*9. Show that every ideal in the polynomial ring $F[x]$, F a field, is a principal ideal.

*10. An *adeal* of a (possibly noncommutative) ring R is a subset M of R such that (a) $x, y \in M$ implies $x + y \in M$ and (b) $x \in M$, $r \in R$ implies $xr \in M$ and $rx \in M$. Construct examples of adeals that are not ideals.

2. Residue Class Rings

Let us first consider a simple example of a residue class ring which, as we will see, we have considered before. To this end we take our ring to be the integers, I. In I we consider the ideal [3] and define, for any $s \in I$, $\bar{s} = \{x | x \in I$ and $s - x \in [3]\}$. Thus

$$\bar{0} = \{\ldots -6, -3, 0, 3, 6, \ldots\}$$
$$\bar{1} = \{\ldots -5, -2, 1, 4, 7, \ldots\}$$
$$\bar{2} = \{\ldots -4, -1, 2, 5, 8, \ldots\}$$

and all elements of S are either in $\bar{0}$, $\bar{1}$, or $\bar{2}$. In fact, $\bar{0}$, $\bar{1}$, and $\bar{2}$ are simply the sets C_0, C_1, and C_2, respectively, defining J_3 as described in Section 5.1. As such they form a ring (in fact, a field) under the addition and multiplication defined previously for J_3.

In general, suppose that we have a ring S and an ideal, R, of S. We define for any $s \in S$,

(1) $$\bar{s} = \{x | x \in S \text{ and } s - x \in R\}$$

Theorem 3. $\bar{s} = \bar{t}$ if and only if $s - t \in R$.

Proof. We note first that $s \in \bar{s}$ since $s - s = 0 \in R$. Now if $\bar{s} = \bar{t}$, $s \in \bar{t}$. Thus $t - s \in R$ and $0 - (t - s) = s - t \in R$.

Conversely, if $s - t \in R$ and $x \in \bar{s}$ we have $s - x \in R$ and hence $(s - x) - (s - t) = t - x \in R$. Thus $x \in \bar{s}$ implies $x \in \bar{t}$. Similarly, if $x \in \bar{t}$ we get $x \in \bar{s}$ and hence have $\bar{s} = \bar{t}$.

We call the sets $\bar{s}, \bar{t}, \ldots$ *residue classes* (modulo R) and make the definitions

(2) $$\bar{s} + \bar{t} = \overline{s + t}$$

and

(3) $$\bar{s} \times \bar{t} = \bar{s}\bar{t} = \overline{st}$$

In terms of our previous example we have

$$\bar{0} + \bar{1} = \overline{0 + 1} = \bar{1}, \bar{1} + \bar{2} = \overline{1 + 2} = \bar{3} = \bar{0} \text{ etc.}$$

and

$$\bar{0} \times \bar{1} = \overline{0 \times 1} = \bar{0}, \bar{1} \times \bar{2} = \overline{1 \times 2} = \bar{2} \text{ etc.}$$

exactly as for the corresponding C_0, C_1, and C_2.

Before we prove that the residue classes form a ring under the addition and multiplication defined by (2) and (3) let us consider another example. Here we will take $S = R^*[x]$, the ring of polynomials in one indeterminate with coefficients real numbers, and R to be the principal ideal $[x^2 + 1]$.

Now any polynomial $f(x) \in S$ may be written as

$$f(x) = (x^2 + 1)g(x) + (ax + b), \; a, b \text{ real numbers}, \; g(x) \in S,$$

so that

$$f(x) - (ax + b) = (x^2 + 1)g(x) \in R$$

Hence, by Theorem 3, $\overline{f(x)} = \overline{ax + b}$ so that we may consider the residue classes of S modulo $[x^2 + 1]$ to be of the form $\overline{ax + b}$. Then, for example, $\overline{x^2 + 1} = \bar{0}$ since $x^2 + 1 = (x^2 + 1)\cdot 1 + 0$. Thus $\overline{x^2} + \bar{1} = \bar{0}$, $\bar{x}^2 + \bar{1} = \bar{0}$, $\bar{x}^2 = -\bar{1} = \overline{-1}$, and

$$\overline{(x + 1)}\,\overline{(x + 2)} = \overline{(x + 1)(x + 2)} = \overline{x^2 + 3x + 2} = \overline{x^2} + \overline{3x + 2}$$

$$= \overline{-1} + \overline{3x + 2}$$

$$= \overline{(-1) + 3x + 2} = \overline{3x + 1}$$

In the following chapter we will consider this particular set of residue classes in a little more detail since it is obvious that it "looks" like the set of complex numbers. We continue now, however, with the general theory.

Theorem 4. The residue classes modulo R of a ring S with ideal R form a ring under addition and multiplication as defined by (2) and (3). We call this ring the *residue class ring* or *difference ring* of S modulo R and write it as $S - R$ or S/R.

Proof. We first need to show that (2) and (3) are adequate definitions of addition and multiplication. That is, we need to show that if $\bar{s} = \overline{s'}$ and $\bar{t} = \overline{t'}$, then

$$(4) \qquad \bar{s} + \bar{t} = \overline{s'} + \overline{t'} \qquad \text{and} \qquad \bar{s}\bar{t} = \overline{s'}\,\overline{t'}$$

Now $\bar{s} + \bar{t} = \overline{s + t}$ and $\overline{s'} + \overline{t'} = \overline{s' + t'}$ and hence we wish to show that $\overline{s + t} = \overline{s' + t'}$. By Theorem 3 the last equality holds if and only if $(s + t) - (s' + t') = (s - s') + (t - t') \in R$. Since $\bar{s} = \overline{s'}$ and $\bar{t} = \overline{t'}$ we have $s - s' \in R$ and $t - t' \in R$. Thus $(s - s') + (t - t') \in R$ as desired since R is closed under addition.

Similarly, we have $\bar{s}\bar{t} = \overline{st}$ and $\overline{s'}\,\overline{t'} = \overline{s't'}$ and need to show that $\overline{st} = \overline{s't'}$. By Theorem 3, the last equality holds if and only if $st - s't' \in R$. Now $s - s' \in R$ and $t - t' \in R$ and thus, since R is an ideal,† $t(s - s') = ts - ts' \in R$ and $s'(t - t') = s't - s't' \in R$. But then, since R is closed under addition, $(ts - ts') + (s't - s't') = st - s't' \in R$ as desired.

† Up to this point in the proof we have used only the fact that R is a subring. Now we need to use the fact that R is an ideal since it does *not* follow that $\overline{st} = \overline{s't'}$ if S is a subring of S but not an ideal of S. See problem 6 of Exercises 9.2.

Once our operations are shown to be well defined it is very easy to verify the ring properties for $S - R$. Thus, for example, $\bar{t} + \bar{s} = \bar{s} + \bar{t}$ since $\bar{t} + \bar{s} = \overline{t + s}$, $\bar{s} + \bar{t} = \overline{s + t}$, and $t + s = s + t$. Observe that $\bar{0}$ is the identity for addition and the additive inverse of $\bar{s}$ is $\overline{-s}$.

A suggestive notation for $\bar{s} = \{x | x \in S \text{ and } s - x \in R\}$ is $s + R$. Then $\bar{s} + \bar{t} = (s + R) + (t + R) = (s + t) + R$ and $\bar{s}\bar{t} = (s + R)(t + R) = st + R$. Furthermore, if $a \in R$, $(s + a) + R = s + R$. In this notation, our first example of J_3 has the elements $\bar{0} = 0 + [3]$, $\bar{1} = 1 + [3]$, $\bar{2} = 2 + [3]$ and our second example has the elements $ax + b + [x^2 + 1]$ where $a, b \in R^*$. Then, for example, $(2 + [3]) + (2 + [3]) = 4 + [3] = (1 + 3) + [3] = 1 + [3]$ since $3 \in [3]$ and $(x + [x^2 + 1])(x + [x^2 + 1]) = x^2 + [x^2 + 1] = -1 + (x^2 + 1) + [x^2 + 1] = -1 + [x^2 + 1]$ since $x^2 + 1 \in [x^2 + 1]$.

Exercises 9.2

1. Prove that the additive inverse of $\bar{s}$ is $\overline{-s}$.
2. Let $S = I$ and $R = [m]$ where m is a positive integer. Prove that $S - R = J_m$.
3. Let $S = I[x]$ and $R = [x - a]$. Prove that $S - R$ is isomorphic to I.
4. Let $S = \{a + bi | a, b \in I\}$ and $R = [2]$. Exhibit $S - R$.
5. Let $S = I[x, y]$ and $R = \{f(x, y) | f(x, y) \in S \text{ and } f(0, 0) = 0\}$. Exhibit $S - R$.
6. Let R and S be defined as in Example 3 of Section 9.1 and consider the residue classes $\bar{x}$ and $\overline{x^3}$. Show that $\bar{x}\,\overline{x^3} = \overline{x^4} = \bar{0}$, $\bar{x} = \overline{x + 1}$, $\overline{x^3} = \overline{x^3 + 1}$ but $\overline{(x + 1)}\,\overline{(x^3 + 1)} \neq \overline{x\,x^3}$.

*7. An ideal R of a ring S is called a *prime* ideal if $ab \in R$ implies $a \in R$ or $b \in R$. Prove that if S has a unity element and R is an ideal of S, then $S - R$ is an integral domain if and only if R is a prime ideal.

3. Homomorphisms

Consider I and $I - [2]$. The mapping that sends even integers into $\bar{0}$ and odd integers into $\bar{1}$ is certainly not an isomorphism of I and $I - [2]$. Nevertheless the correspondence is preserved under addition and multiplication since

$$\begin{aligned}
\text{odd} + \text{odd} &= \text{even} \rightarrow \bar{1} + \bar{1} = \bar{0} \\
\text{odd} + \text{even} &= \text{odd} \rightarrow \bar{1} + \bar{0} = \bar{1} \\
\text{even} + \text{even} &= \text{even} \rightarrow \bar{0} + \bar{0} = \bar{0} \\
\text{odd} \times \text{odd} &= \text{odd} \rightarrow \bar{1} \times \bar{1} = \bar{1} \\
\text{odd} \times \text{even} &= \text{even} \rightarrow \bar{1} \times \bar{0} = \bar{0} \\
\text{even} \times \text{even} &= \text{even} \rightarrow \bar{0} \times \bar{0} = \bar{0}.
\end{aligned}$$

Definition 4. *A **homomorphism** of a ring S into a ring S' is a mapping $\sigma : S \to S'$ with the property that the correspondence is preserved under addition and multiplication; that is, if $s, t \in S$, then $(s + t)\sigma = s\sigma + t\sigma$ and $(st)\sigma = (s\sigma)(t\sigma)$. If for every $s' \in S'$ there exists an $s \in S$ such that $s\sigma = s'$ we say that the homomorphism is **onto**.*

If there exists a homomorphism of S into S' we will say that S is *homomorphic* to S' and call S' the *homomorphic image* of S. An *isomorphism* of S and S' is a homomorphism that is 1–1 and onto.

The example given above of a homomorphism of I onto $I - [2]$ can obviously be generalized to a homomorphism of I onto $I - [n]$ for any positive integer n. Thus if $n_1 \to \bar{n}_1$, $n_2 \to \bar{n}_2$, then

$$n_1 + n_2 \to \overline{n_1 + n_2} = \bar{n}_1 + \bar{n}_2 \qquad \text{and} \qquad n_1 n_2 \to \overline{n_1 n_2} = \bar{n}_1 \bar{n}_2,$$

i.e., if $n\sigma = n$ we have

$$(n_1 + n_2)\sigma = \overline{n_1 + n_2} = \bar{n}_1 + \bar{n}_2 = n_1\sigma + n_2\sigma \qquad \text{and}$$

$$(n_1 n_2)\,\sigma = \overline{n_1 n_2} = \bar{n}_1 \bar{n}_2 = (n_1\sigma)(n_2\sigma)$$

An isomorphism is a homomorphism that is 1–1 and onto. In general, a homomorphism is a many-one mapping and is not onto. Thus in the homomorphism of I into $I - [2]$ we have $-2 \to \bar{0}$, $0 \to \bar{0}$, $2 \to \bar{0}$ and so on, although, in this case the mapping is obviously onto.

On the other hand, if $S = I[x]$ and $S' = \{a + b\sqrt{2} | a, b \in R\}$, the mapping σ defined by $f(x)\sigma = f(\sqrt{2})$ for $f(x) \in I[x]$ is a homomorphism that is neither 1–1 nor onto. Thus, for example, $(2x)\sigma = 2\sqrt{2} = x^3\sigma$ so that the mapping is not 1–1. Furthermore, if we consider $\frac{1}{2} + \sqrt{2} \in S'$ it is clear that there is no $f(x) \in I[x]$ such that $f(x)\sigma = \frac{1}{2} + \sqrt{2}$. However, the mapping is a homomorphism since for $f(x), g(x) \in I[x]$, we have

$$f(x) + g(x) = h(x) \in I[x], \qquad f(x)g(x) = k(x) \in I[x],$$

and

$$[f(x) + g(x)]\sigma = h(x)\sigma = h(\sqrt{2}) = f(\sqrt{2}) + g(\sqrt{2}) = f(x)\sigma + g(x)\sigma;$$

$$[f(x)g(x)]\sigma = k(x)\sigma = k(\sqrt{2}) = f(\sqrt{2})g(\sqrt{2}) = [f(x)\sigma][g(x)\sigma]$$

As a final example of a homomorphism consider the mapping, σ, of I into R defined by $n\sigma = n$ for $n \in I$. Then σ is obviously a homomorphism which is 1–1 but not onto. On the other hand, the mapping, τ, of I into R defined by $n\tau = 2n$ for $n \in I$ is not a homomorphism. We do have $(n_1 + n_2)\tau = 2(n_1 + n_2) = 2n_1 + 2n_2 = n_1\tau + n_2\tau$ but $(n_1 n_2)\tau = 2(n_1 n_2) \neq (2n_1)(2n_2) = (n_1\tau)(n_2\tau)$ (unless $n_1 n_2 = 0$).

Theorem 5. Suppose that the ring S is homomorphic to the ring S' under the homomorphism σ. Then if 0 is the additive identity for S, we have $0\sigma = 0'$, the additive identity for S', and if $s\sigma = s'$ for $s \in S$, we have $(-s)\sigma = -s'$.

Proof. (1) Let $0\sigma = x'$ and let s' be any element of S'. Then

$$x' + s' = 0\sigma + s' = (0 + 0)\sigma + s' = (0\sigma + 0\sigma) + s' = x' + (x' + s').$$

Thus $s' = x' + s'$ for all $s' \in S'$ and hence x' is the additive identity for S'.

(2) We have $[s + (-s)]\sigma = s\sigma + (-s)\sigma$. But $[s + (-s)]\sigma = 0\sigma = 0' \in S'$ by (1). Hence $0' = s\sigma + (-s)\sigma = s' + (-s)\sigma$ and hence $(-s)\sigma = -s'$.

We leave as an exercise for the student the proof of the following corollary.

Corollary. For all $s, r \in S$, $(s - r)\sigma = s\sigma - r\sigma$.

We also leave as an exercise for the student to show that commutativity of addition in S and S' is not vital to the establishment of our conclusions in Theorem 5—a result which will find later application in considering homomorphisms of groups.

Note that we do *not* have a corresponding result to Theorem 5 for multiplication even when S and S' both have multiplicative identities. This is illustrated by the homomorphism σ of I into $I - [2]$ defined by $n\sigma = \bar{0}$ for all $n \in I$. We do have $0\sigma = \bar{0}$ as required by Theorem 5, but $1\sigma = \bar{0} \neq \bar{1}$, the multiplicative identity of $I - [2]$. We leave it as an exercise for the student to prove that if the homomorphism is *onto*, the corresponding result *does* hold for multiplication when S has a multiplicative identity.

We have seen that if σ is a homomorphism of a ring S into a ring S', then at least one element of S, namely 0, maps into the additive identity, $0'$, of S'. Of special interest is the set of *all* the elements of S which map into $0'$.

Definition 5. *Let σ be a homomorphism of a ring S into a ring S' and let $0'$ be the additive identity of S'. Then*

$$K_\sigma = \{s \mid s \in S \text{ and } s\sigma = 0'\}$$

is called the ***kernel*** *of the homomorphism σ.*

Ideals and homomorphisms are closely related. If S and S' are rings, every homomorphism of S onto S' determines an ideal of S and, conversely, every ideal of S determines a ring S' and a homomorphism of S onto S'. How the determinations are made is given explicitly in the following theorem.

Theorem 6. (1) Suppose that the ring S is homomorphic to the ring S' under the onto homomorphism σ and let $0'$ be the additive identity of S'. Then K_σ, the kernel of S under σ, is an ideal of S. Furthermore, $S - K_\sigma \cong S'$.

(2) Let R be an ideal of a ring S and for $s \in S$ define $\bar{s} = \{x | x \in S$ and $s - x \in R\} = s + R$. Then $s\sigma = \bar{s}$ defines a homomorphism, σ, of S onto $S - R$ with kernel R.

Before proving the theorem let us consider an example. First suppose that $S = I$ and $S' = J_3 = \{C_0, C_1, C_2\}$. Our homomorphism is defined by

$$(3n)\sigma = C_0, (3n + 1)\sigma = C_1, \text{ and } (3n + 2)\sigma = C_2$$

Then $K_\sigma = \{3n | n \in I\} = [3]$. Our isomorphism of $S - K_\sigma$ and S' is defined by $n + K_\sigma \leftrightarrow C_n$ for $n = 0, 1, 2$.

On the other hand, suppose $S = I$ and let $R = [3]$. Then, as we have seen, $I - [3] = \{\bar{0}, \bar{1}, \bar{2}\}$ and our homomorphism, σ, is defined by $(3n)\sigma = \bar{0}$, $(3n + 1)\sigma = \bar{1}$, and $(3n + 2)\sigma = \bar{2}$.

We now turn to a proof of our theorem.

(1) We use Lemma 1 to show that K_σ is an ideal. Thus if $k_1, k_2 \in K_\sigma$, we have $k_1\sigma = 0'$ and $k_2\sigma = 0'$. But then $(k_1 - k_2)\sigma = k_1\sigma - k_2\sigma = 0' - 0' = 0'$ by the corollary to Theorem 5 so that $k_1 - k_2 \in K_\sigma$. Also, if $k \in K_\sigma$ and $s \in S$, $k\sigma = 0'$, $s\sigma = s'$, we have $(sk)\sigma = (s\sigma)(k\sigma) = s' \cdot 0' = 0'$ so that $sk \in K_\sigma$, $SK_\sigma \subseteq K_\sigma$, and K_σ is an ideal of S. Our isomorphism of $S - K_\sigma$ and S' is the mapping τ defined by $\bar{s}\tau = s'$ where $\bar{s} = s + K_\sigma \in S'$ and s' is determined by $s\sigma = s'$. Since σ is an onto mapping it is clear that τ is also. To show that τ is a 1–1 mapping suppose that $\bar{s}\tau = s'$ and also $\bar{r}\tau = s'$. Then $s\sigma = s'$ and $r\sigma = s'$. Hence, by the corollary to Theorem 5, $s\sigma - r\sigma = (s - r)\sigma = s' - s' = 0'$. But then $s - r \in K_\sigma$ so that, by Theorem 3, $\bar{s} = \bar{r}$. Finally, that τ is preserved under addition and multiplication is shown by observing that

$$(\bar{s} + \bar{t})\tau = (\overline{s + t})\tau = (s + t)' = (s + t)\sigma = s\sigma + t\sigma = s' + t' = \bar{s}\tau + \bar{t}\tau$$

and

$$(\bar{s}\bar{t})\tau = (\overline{st})\tau = (st)' = (st)\sigma = (s\sigma)(t\sigma) = s't' = (\bar{s}\tau)(\bar{t}\tau)$$

where we are using the definitions for $\bar{s} + \bar{t}$ and $\bar{s}\bar{t}$ given following Theorem 3.

(2) The mapping described is an onto mapping since every $\bar{s}$ in $S - R$ is equal to $s\sigma$ for $s \in S$. Furthermore, $(s + r)\sigma = \overline{s + r} = \bar{s} + \bar{r} = s\sigma + r\sigma$ and $(sr)\sigma = \overline{sr} = \bar{s}\bar{r} = (s\sigma)(r\sigma)$ so that the mapping is a homomorphism. The kernel of the homomorphism consists of those elements $s \in S$ such that $s\sigma = \bar{s} = \bar{0}$; i.e., those elements $s \in S$ such that $s \in R$.

As a second example of our theorem, consider the homomorphism σ defined by $f(x)\sigma = f(\sqrt{2})$ where $f(x) \in S = I[x]$ and $f(\sqrt{2}) \in S' = \{a + b\sqrt{2} | a, b \in I\}$. If $f(x)\sigma = 0' = 0 + 0 \cdot \sqrt{2} \in S'$, then $f(\sqrt{2}) = 0$. The ideal $K_\sigma = \{f(x) | f(x) \in I[x]$ and $f(\sqrt{2}) = 0\}$ of $I[x]$ is the kernel of the homomorphism. We have (problem 4 of Exercises 9.1) that $K_\sigma = [x^2 - 2]$.

On the other hand, if we let $S = I[x]$ and take $R = [x^2 - 2]$ we have $S - R = \{s + R | s \in S\}$. Our homomorphism τ of S onto $S - R$ is defined by writing any $f(x) \in S$ as $f(x) = g(x)(x^2 - 2) + ax + b$ and having $f(x)\tau = (ax + b) + R$. Then $S - R \cong \{a + b\sqrt{2} | a, b \in I\}$ via $(ax + b) + R \leftrightarrow a + b\sqrt{2}$. (Cf. $R^*[x] - [x^2 + 1] \cong C$ as suggested in Section 2.)

Exercises 9.3

1. Which of the following mappings are homomorphisms? If the mapping is a homomorphism, determine the kernel of the homomorphism.
   (a) $S = R[x]$, $S' = \{a + bi | a, b \in R\}$, $f(x)\sigma = f(i)$ for $f(x) \in S$;
   (b) $S = I$, $S' = \{n^2 + 2 | n \in I\}$, $s\sigma = s^2 + 2$ for $s \in S$;
   (c) $S = R[x]$, $S' = \{a + bi | a, b \in R\}$, $f(x)\sigma = f(\omega)$ for $f(x) \in S$ where ω is the complex, nonreal cube root of unity, $\frac{1}{2} + (\sqrt{3}/2)i$.
   (d) Same as (c) except that R is replaced by R^*;
   (e) $S = I$, $S' = J_4$, $(2n)\sigma = C_0$, $(2n + 1)\sigma = C_2$ for $n \in I$;
   (f) $S = J_4$, $S' = J_2$, $C_0\sigma = C'_0 \in S'$, $C_1\sigma = C'_1 \in S'$, $C_2\sigma = C'_0 \in S'$, $C_3\sigma = C'_1 \in S'$.
2. Prove that the homomorphic image of a commutative ring is a commutative ring.
3. What are the possible homomorphic images of a field?
4. Prove the corollary to Theorem 5.
5. Show that the result of Theorem 5 still holds if addition in S and S' is noncommutative.
6. Show that if S is a ring with multiplicative identity, e, and the mapping $\sigma : S \to S'$ is a homomorphism *onto* the ring S', then $e\sigma$ is the multiplicative identity for S'.
7. (a) Suppose that the homomorphism σ of S into S' is an isomorphism. What is the kernel of σ in this situation?
   (b) Suppose $S' = \{0\}$. What is the kernel of S?
8. (a) Suppose that the ideal, R, of S is equal to S. What is $S - R$?
   (b) Suppose $R = \{0\}$. What is $S - R$?
9. Find all of the homomorphic images of J_{10}. (*Hint.* Use the results of problem 3 (*b*) of Exercises 9.1 together with part (1) of Theorem 6.)

4. Homomorphisms of Groups

We begin by rephrasing our definitions of homomorphisms of rings in terms of groups.

Definition 6. *Let G and G' be groups with operations $\circ$ and $*$ respectively. Then a mapping, σ, of G into G' is called a* **homomorphism** *of G into G' if, for all $a, b \in G$, $(a \circ b)\sigma = (a\sigma) * (b\sigma)$. As before, we customarily write $(ab)\sigma = (a\sigma)(b\sigma)$. We say that G is* **homomorphic** *to G' and call G' the* **homomorphic image** *of G. The set $K_\sigma = \{g | g \in G \text{ and } g\sigma = e'\}$*

where e' is the identity of G' is called the **kernel** *of the homomorphism.*

Theorem 7. If σ is a homomorphism of G into G', e is the identity of G, and e' is the identity of G', then $e\sigma = e'$ and if $a\sigma = a'$ for $a \in G$, then $(a^{-1})\sigma = (a')^{-1}$.

We leave the proof as an exercise for the student.

Example 1. Let $G = S_n$ be the symmetric group on n symbols (Section 7.4) and let $G' = \{1, -1\}$ under multiplication. If p is an even permutation of G we let $p\sigma = 1$ and if q is an odd permutation of G we let $q\sigma = -1$. It is easy to see that σ is a homomorphism of G onto G' with kernel the alternating group, A_n, of even permutations on n symbols. (See problem 6 of Exercises 7.4.)

Example 2. A homomorphism σ of the alternating group, A_4, on four symbols onto the cyclic group of order 3 may be defined as follows:

$$e\sigma = [(1)(2)(3)(4)]\sigma = e' = a^3$$

$$[(12)(34)]\sigma = e'$$

$$[(13)(24)]\sigma = e'$$

$$[(14)(23)]\sigma = e'$$

$$(123)\sigma = a \qquad (132)\sigma = a^2$$

$$(243)\sigma = a \qquad (234)\sigma = a^2$$

$$(142)\sigma = a \qquad (124)\sigma = a^2$$

$$(134)\sigma = a \qquad (143)\sigma = a^2$$

The kernel of σ is $\{e, (12)(34), (13)(24), (14)(23)\}$.

The proof of the following theorem is left to the student.

Theorem 8. If σ is a homomorphism of a group G into a group G', then the kernel of σ is a subgroup of G.

Examples 1 and 2 illustrate Theorem 8.

In our work with rings we have seen (Theorem 6) that every homomorphism of a ring S onto a ring S' determines an ideal of S (not just a subring of S) and that, conversely, every ideal of S (not just a subring of S) determines a homomorphism of S onto a ring S'. By analogy, then, we might expect that for groups a homomorphism determines a special kind of subgroup and, conversely, that a special kind of subgroup is needed to determine a homomorphism.

Now the key concept in establishing a homomorphism of a ring was that of the residue classes, $s + R = \{x | x \in S \text{ and } s - x \in R\}$. But if $s - x = r \in R$, then $x = s + (-r) = x + r_1$, where $r_1 \in R$. Conversely, if $x = s + r_1$ where

$r_1 \in R$, then $s - x = -r_1 = r \in R$. Thus $s + R = \{x | x \in S \text{ and } x = s + r \text{ for } r \in R\}$. Shifting from the additive notation to the multiplicative notation we define the analogous concept for groups.

Definition 7. *Let H be a subgroup of a group G and $g \in G$. Then we define the* **left coset,** *gH, of H in G by $gH = \{gh | h \in H\}$. Similarly, the* **right coset,** *Hg, is defined by $Hg = \{hg | h \in H\}$.*

Example 3. Let G be the *octic* permutation group whose elements are $i = (1)(2)(3)(4)$, $a = (1234)$, $b = (13)(24)$, $c = (1432)$, $d = (13)$, $e = (24)$, $f = (12)(34)$, $g = (14)(23)$, and let $H = \{i, d\}$. Then the left cosets of H in G are:

$$iH = \{ii, id\} = \{i, d\}; \qquad aH = \{ai, ad\} = \{a, f\};$$

$$dH = \{di, dd\} = \{d, i\}; \qquad fH = \{fi, fd\} = \{f, a\};$$

$$bH = \{bi, bd\} = \{b, e\}; \qquad cH = \{ci, cd\} = \{c, g\};$$

$$eH = \{ei, ed\} = \{e, b\}; \qquad gH = \{gi, gd\} = \{g, c\}$$

Note that $H = iH = dH$; $aH = fH$; $bH = eH$; $cH = gH$. Thus if we have two equal cosets xH and yH it does not follow that $x = y$ but only that for every $h \in H$ there exists $h' \in H$ such that $xh = yh'$.

When the group is Abelian we customarily use the additive notation as for rings and write $g + H$ for gH. Notice that, in the Abelian case, $g + H = H + g$ since $g + h = h + g$ for all $g \in G$ and $h \in H$.

Example 4. Let G be the additive group of integers and $H = \{3n | n \in I\}$. Then $H = 0 + H = H + 0$; $1 + H = H + 1 = \{3n + 1 | n \in I\}$; and $2 + H = H + 2 = \{3n + 2 | n \in I\}$. Note that for any integer n, $3n + H = H$, $(3n + 1) + H = 1 + H$, and $(3n + 2) + H = 2 + H$.

Now in our discussion of rings, after we had defined the residue classes, $\bar{s} = s + R$, we proceeded to define addition and multiplication of these residue classes by $\bar{s} + \bar{t} = (s + R) + (t + R) = \overline{s + t} = (s + t) + R$ and $\bar{s}\bar{t} = (s + R)(t + R) = \overline{st} = st + R$. Before, however, we could proceed to show that the residue classes formed a ring we needed to show (Theorem 4) that if $s + R = s' + R$ and $t + R = t' + R$, then $(s + R) + (t + R) = (s' + R) + (t' + R)$ and $(s + R)(t + R) = (s' + R)(t' + R)$.

Similarly, in the case of groups we wish to define $(aH)(bH) = (ab)H$ and need to prove that if $aH = a'H$ and $bH = b'H$, then $(aH)(bH) = (a'H)(b'H)$. As a matter of fact, however, no such result can be established. Consider in Example 3, aH and cH. Then $(aH)(cH) = (ac)H = iH$. However, $aH = fH$ $cH = gH$, and $(fH)(gH) = bH \neq iH$. Thus, as we rather expected, the desired result is not true for arbitrary cosets; there must be a restriction.

To establish a theorem concerning this restriction it is useful to prove the following lemma.

Lemma 2. If H is any subgroup of a group G, then for any $g \in G$ and $h \in H$, $gH = (gh)H$.

Proof. Let e be the identity for G. Then for any $gh \in gH$ we have $gh = (gh)e \in (gh)H$ so that $gH \subseteq (gh)H$. On the other hand, for any $(gh)h_1 \in (gh)H$ we have $(gh)h_1 = g(hh_1) = gh_3 \in gH$ since H is a subgroup of G. Hence $(gh)H \subseteq gH$ so that $gH = (gh)H$.

Theorem 9. Let G be a group, H a subgroup of G, and define $(aH)(bH) = (ab)H$. If $aH = a'H$ and $bH = b'H$ implies $(ab)H = (a'b')H$, then $gH = Hg$ for all $g \in G$. Conversely, if $gH = Hg$ for all $g \in G$, then $aH = a'H$ and $bH = b'H$ implies $(ab)H = (a'b')H$.

Proof. Suppose that $(ab)H = (a'b')H$ whenever $aH = a'H$ and $bH = b'H$. For any $g \in G$ we have $(gH)(g^{-1}H) = (gg^{-1})H = eH = H$ where e is the identity element of G. Now by our lemma, $gH = (gh)H$ for $h \in H$, so that our hypothesis demands that $[(gh)H][g^{-1}H] = (ghg^{-1})H = H$ for any $h \in H$. Thus for any $h \in H$ there exists $h_1 \in H$ such that $(ghg^{-1})e = ghg^{-1} = h_1$ so that $gh = h_1 g$. Hence $gH \subseteq Hg$. Similarly, by considering $[(g^{-1}h)H][gH]$, we may prove that $Hg \subseteq gH$ and hence conclude that $gH = Hg$.

Now suppose that $gH = Hg$ for all $g \in G$ and that $aH = a'H$ and $bH = b'H$. Now any element of $(ab)H$ is of the form $(ab)h$ for $h \in H$, and we know that $ae = a = a'h_1$ and $be = b = b'h_2$ for some $h_1, h_2 \in H$. Then $(ab)h = (a'h_1)(b'h_2)h = a'(h_1b')(h_2h) = a'(b'h_3)(h_2h)$ for $h_3 \in H$ since we are supposing that $gH = Hg$. Thus $(ab)h = (a'b')(h_3h_2h) = (a'b')h_4$ for $h_4 \in H$ since H is a subgroup of G. Hence $(ab)H \subseteq (a'b')H$. Similarly, we obtain $(a'b')H \subseteq (ab)H$ and conclude that $(ab)H = (a'b')H$.

For an alternative approach to the multiplication of cosets of a subgroup H of G and the restrictions on H see problem 18 of Exercises 9.4.

Definition 8. *If H is a subgroup of a group G such that $gH = Hg$ for all $g \in G$, we call H a* **normal** *subgroup of G; notation: $H \Delta G$.*

Example 5. In Example 3, the subgroup $H = \{i, d\}$ is not a normal subgroup. We do have $iH = Hi$, $dH = Hd$ but, for example, $Ha = \{a, g\} \neq aH = \{a, f\}$. On the other hand, the subgroup $S = \{i, b\}$ is a normal subgroup. We have

$$iS = bS = \{i, b\} = Si = Sb;$$
$$aS = cS = \{a, c\} = Sa = Sc;$$
$$dS = eS = \{d, e\} = Sd = Se;$$
$$fS = gS = \{f, g\} = Sf = Sg$$

Since it is easy to show that $gH = Hg$ if and only if $gHg^{-1} = H$, a normal subgroup is sometimes called a *self conjugate* subgroup by analogy with the definition of gag^{-1} as the conjugate of a under g. (See Definition 7.14.)

Theorem 10. Let H be a subgroup of a group G. Then the left cosets, aH, of H in G form a group under the multiplication $(aH)(bH) = (ab)H$ if and only if $H \Delta G$. This group is called the *quotient* or *factor* group of G modulo H and is written G/H or, for Abelian groups, as $G - H$.

Proof. The necessity of the condition has been established by Theorem 9 and this theorem also tells us that, if $H \Delta G$, the multiplication of cosets, $(aH)(bH) = (ab)H$, is well defined. It is obvious that this multiplication is associative; the identity element of G/H is eH where e is the identity for G; and the inverse of aH is $a^{-1}H$. Thus if $H \Delta G$, G/H is a group.

Example 6. Consider the cosets of $S = \{i, b\}$ in the octic group G (Example 5). We have $G/S = \{iS, aS, dS, fS\}$ with the following multiplication table:

	iS	aS	dS	fS
iS	iS	aS	dS	fS
aS	aS	iS	fS	dS
dS	dS	fS	iS	aS
fS	fS	dS	aS	iS

We now prove the analog of Theorem 6 for groups.

Theorem 11. (1) Suppose σ is a homomorphism of G onto G'. Then if K_σ is the kernel of G, we have $K_\sigma \Delta G$ and $G/K_\sigma \cong G'$.

(2) Suppose $H \Delta G$. Then $g\sigma = gH$ defines a homomorphism, σ, of G onto G/H with kernel H. (The order of G/H is called the *index* of H in G.)

Proof. (1) We have already shown (Theorem 8) that K_σ is a subgroup of G. Now suppose that g is any element of G and k any element of K_σ. By Theorem 7, we have $(gkg^{-1})\sigma = (g\sigma)(k\sigma)(g^{-1}\sigma) = g'e'(g')^{-1} = e'$ so that $gkg^{-1} \in K_\sigma$. Thus for any $g \in G$ and any $k \in K_\sigma$ we know that there exists $k_1 \in K_\sigma$ such that $gkg^{-1} = k_1$ or $gk = k_1g$. But this means that $gK_\sigma \subseteq K_\sigma g$. Using $g^{-1}kg$ in place of gkg^{-1} we obtain $K_\sigma g \subseteq gK_\sigma$ and hence conclude that $gK_\sigma = K_\sigma g$ for all $g \in G$ so that $K_\sigma \Delta G$.

Our isomorphism, τ, of G/K_σ and G' is defined by $(aK_\sigma)\tau = a'$, for $aK_\sigma \in G/K_\sigma$ and $a' \in G'$, if and only if $a\sigma = a'$. Since σ is an onto mapping it is clear that τ is also. To show that τ is a 1–1 mapping suppose that $(aK_\sigma)\tau = a'$ and also $(bK_\sigma)\tau = a'$. Then $a\sigma = a'$ and $b\sigma = a'$ so that $(b^{-1}a)\sigma = (b^{-1}\sigma)(a\sigma) = (a')^{-1}a' = e'$. Hence $b^{-1}a = k \in K_\sigma$. But then

$aK_\sigma = (bk)K_\sigma = bK_\sigma$ by Lemma 2. Finally, the mapping τ is preserved under the group operation since $[(aK_\sigma)(bK_\sigma)]\tau = [(ab)K_\sigma]\tau = (ab)' = (ab)\sigma = (a\sigma)(b\sigma) = a'b' = [(aK_\sigma)\tau][(bK_\sigma)\tau]$.

(2) Since $(ab)\sigma = (ab)H = (aH)(bH) = (a\sigma)(b\sigma)$ and since every element $gH \in G/H$ is the image of some element (for example, g) of G, the mapping σ is a homomorphism of G onto G/H. Since the identity of G/H is eH where e is the identity of G, the kernel of the homomorphism consists precisely of those elements of G such that $gH = eH = H$. Since it is easy to see that $gH = H$ if and only if $g \in H$, it follows that the kernel of the homomorphism is H.

We leave the proof of the following corollary as an exercise for the student.

Corollary. A homomorphism σ of a group G into a group G' with kernel K_σ is an isomorphism of G onto G' if and only if $K_\sigma = \{e\}$ where e is the identity of G.

Example 7. (1) Let G be the octic group $\{i, a, b, c, d, e, f, g\}$ as defined in Example 3 and let G' be the group $\{e, x, y, z\}$ with multiplication table

	e	x	y	z
e	e	x	y	z
x	x	e	z	y
y	y	z	e	x
z	z	y	x	e

It may be checked that G is homomorphic to G' via the mapping σ: $i, b \to e$; $a, c \to x$; $d, e \to y$; $f, g \to z$. The kernel is $K_\sigma = \{i, b\}$ which is, as we have seen in Example 5, a normal subgroup of G. Furthermore, $G/K_\sigma \cong G'$ under the 1–1 onto mapping

$$iK_\sigma \leftrightarrow e; \quad aK_\sigma \leftrightarrow x; \quad dK_\sigma \leftrightarrow y; \quad fK_\sigma \leftrightarrow z.$$

(2) Using the results of Example 6, with G as the octic group and $H = \{i, b\}$, we have $i, b \to iH$; $a, c \to aH$; $d, e \to dH$; $f, g \to fH$ defining our homomorphism of G onto G/H. The elements i and b map into iH so that the kernel of the homomorphism is $\{i, b\} = H$.

Theorem 11 tells us, in effect, that the problem of finding all of the homomorphisms of a group is equivalent to the problem of finding all of the normal subgroups of a group. In trying to determine the subgroups of a finite group, a powerful tool is the fact that the order of a subgroup of a group must divide the order of the group. We will establish this fact in the next section; in the meantime, however, it may be used to advantage in some of the problems given in the following set of exercises.

Exercises 9.4

1. Let G be the octic group (Example 3) and let $S = \{i, e\}$. List the right cosets of S in G.
2. Let G be the symmetric group on four symbols and S be the octic group. List the left cosets of S in G.
3. Let G be the alternating group on four symbols and $S = \{(1)(2)(3)(4), (12)(34), (13)(24), (14)(23)\}$. List the left cosets of S in G.
4. Determine all of the normal subgroups of the group of symmetries of the square (Section 7.9) and determine the quotient group for each of these normal subgroups.
5. Determine all of the normal subgroups of the symmetric group on three symbols and determine the quotient group for each of these normal subgroups.
6. Set up a homomorphism of the additive group of integers onto the multiplicative group $\{1, -1, i, -i\}$ $(i = \sqrt{-1})$. What is the kernel of the homomorphism?
7. Which of the following mappings map the multiplicative group of all nonzero real numbers homomorphically on a part of itself? If the mapping is a homomorphism, identify the homomorphic image and the kernel of the homomorphism.
   (a) $x \to -x$; (b) $x \to |x|$; (c) $x \to 3x$; (d) $x \to x^2$;
   (e) $x \to x^3$; (f) $x \to 1/x$; (g) $x \to -1/x$; (h) $x \to \sqrt[3]{x}$.
8. Set up a homomorphism of the octic group onto the group $\{(1)(2)(3)(4), (12)(34), (13)(24), (14)(23)\}$. What is the kernel of the homomorphism? (*Hint:* See Example 7.)
9. Set up a homomorphism of the octic group onto the cyclic group of order two. What is the kernel of the homomorphism? (*Hint:* Find a normal subgroup of order four of the octic group.)
10. In problem 4, find the homomorphisms determined by the normal subgroups.
11. Do as in problem 10 for problem 5.
12. If H is a subgroup of a group G, prove that for $g \in G$, $gH = H$ if and only if $g \in H$.
13. Prove Theorem 7.
14. Prove Theorem 8.
15. Prove the corollary to Theorem 11.
16. Prove that the homomorphic image of a cyclic group is cyclic.
17. Prove that the center S (problem 4, Exercises 7.9) of a group G is a normal subgroup of G and that G/S is isomorphic to the group of inner automorphisms of G.

*18. If S_1 and S_2 are two subsets of a group G, define their *product*, $S_1 * S_2$, as $S_1 * S_2 = \{s_1s_2 | s_1 \in S_1, s_2 \in S_2\}$. Then (1) prove that if H is a normal subgroup of G, then $(aH) * (bH) = (ab)H$ for all $a, b \in G$; (2) if $(aH) * (bH)$ is a left coset of S in G for all $a, b \in G$, then H is a normal subgroup of G.

*19. Let G be a group. For $a, b \in G$ we define the *commutator* of a and b as $a^{-1}b^{-1}ab$. Let S be the set of all products of such commutators; prove that S is a normal subgroup of G.

*20. In problem 17, prove that G/S is Abelian.

*21. Two subgroups S and T of a group G are called *conjugate* if there exists a $g \in G$ such that $g^{-1}Sg = T$. Given any subgroup S of G prove that the intersection of S with all of its conjugates is a normal subgroup of G.

*22. In problem 4 of Exercises 7.9 the center of a group G was defined. Now define the *rim*, $R(G)$ of a group as $R(G) = \{a \in G | ab = ba$ implies there exists $c \in G$ such that $a = c^j$, $b = c^k$ for integers j and $k\}$ and the *anti-center* of G, $AC(G)$, as $AC(G) = \{a_1a_2 \ldots a_n | a_i \in R(G)\}$. Prove:
(a) If e is the identity of G, then $e \in R(G)$;
(b) $a \in R(G)$ implies $a^{-1} \in R(G)$;
(c) $a \in R(G)$ and $b \in G$ implies $b^{-1}ab \in R(G)$;
(d) $AC(G)$ is a normal subgroup of G;
(e) If G is cyclic, then $R(G) = AC(G) = G$;
(f) If G is finite and Abelian, then $R(G) = AC(G) = G$ if and only if G is cyclic;
(g) $G \cong G^*$ implies $AC(G) \cong AC(G^*)$;
(h) $AC(G) = AC(AC(G))$.

5. Lagrange's Theorem

In this section we continue to work with left cosets but the student should note that our results, here and in the previous section, could equally well be phrased in terms of right cosets. For normal subgroups, of course, the concepts are identical since if H is normal, then $gH = Hg$.

Definition 9. *If S is a subgroup of a group G we will say that $a \in G$ is* ***left congruent*** *to $b \in G$ if and only if $a \in bS$. We write $a \equiv b$ (mod S).*

If we let G be the group of integers under addition and $S = \{3n | n \in I\}$, the cosets of S are $0 + S = S$, $1 + S = \{3n + 1 | n \in I\}$, and $2 + S = \{3n + 2 | n \in I\}$ (using, of course, the additive notation). Then to say that $a \equiv b$ (mod S) means that $a \in b + S$ or that $a = b + s = b + 3n$ or that $a - b = 3n$. Thus $a \equiv b$ (mod S) means the same as our previous $a \equiv b$ (mod 3) and this fact helps to explain the language used in Definition 9.

Lemma 3. The relation of left congruence is an equivalence relation on G.

Proof. (1) $a \equiv a$ (mod S) since $a = ae \in aS$; (2) If $a \equiv b$ (mod S) then $a = bs \in bS$. Thus $b = as^{-1} \in aS$ so that $b \equiv a$ (mod S); (3) If $a \equiv b$ (mod S) and $b \equiv c$ (mod S), then $a = bs_1 \in bS$ and $b = cs_2 \in cS$. Hence $a = (cs_2)s_1 = c(s_2s_1) = cs_3 \in S$ and $a \equiv c$ (mod S).

We may now apply Theorem 4.1 to conclude that a group G is partitioned by the classes $P_b = \{a | a \in G$ and $a \in bS\} = bS$. In particular, then, every element of G is in one and only one coset of S. Thus if aS and bS are two cosets, either $aS = bS$ or $aS \cap bS = \varnothing$. Furthermore, if G is finite, the number of elements in any two cosets is the same and is equal to the order of S.

We state a part of the results just obtained as well as the corresponding results for right cosets as a theorem.

Theorem 12. If aS and bS (Sa and Sb) are two left (right) cosets, then either $aS = bS(Sa = Sb)$ or $aS \cap bS = \varnothing (Sa \cap Sb = \varnothing)$.

Theorem 13. (Lagrange's theorem) The order of a subgroup S of a finite group G is a divisor of the order of G.

Proof. Let n be the order of G and s the order of S. Separate the elements of G into left cosets. By our previous remarks, every element of G is in one and only one left coset and each left coset contains the same number of elements, say k. Hence $n = sk$ and $s|n$.

Corollary 1. The order of any element of a group of finite order divides the order of the group.

Proof. Any element of a group generates a cyclic subgroup of the group whose order is the order of the element.

Corollary 2. Every group of prime order is cyclic.

Proof. Let G be a group of prime order, p, with identity e. If $a \in G$ and $a \neq e$, the order of the cyclic group generated by a must divide p and hence is of order p itself. Thus G consists of the powers of a.

Corollary 3. (Fermat's theorem) If $a \in I$ and p is a prime, then $a^p \equiv a \pmod p$.

Proof. The multiplicative group of nonzero elements of J_p is of order $p - 1$ with $\bar{1}$ as its identity. Thus if $a \not\equiv 0 \pmod p$, $a^{p-1} \equiv 1 \pmod p$ and hence $a^p \equiv a \pmod p$. If $a \equiv 0 \pmod p$, the result is obvious.

Exercises 9.5

1. Check Fermat's theorem for $p = 5$ and $a = 2, 3, 4$.
2. Prove that the number of left cosets of a subgroup of any finite group is equal to the number of right cosets.
3. Let G be a group, S a subgroup of G, and $x, y \in G$. Define $x \equiv y \pmod S$ if and only if $x = ys$ for $s \in S$. Prove that if $x \equiv y \pmod S$, then $ax \equiv ay \pmod S$ for all $a \in G$. On the other hand, prove that if $x \equiv y \pmod S$, then $xa \equiv ya \pmod S$ for all $a \in G$ if and only if $S \Delta G$.
4. Prove that a group of order p^m where p is a prime must contain a subgroup of order p.
5. Determine (up to an isomorphism) all groups of order 4.
***6.** Do as in problem 5 for groups of order 6.

PROBLEMS FOR FURTHER INVESTIGATION

1. Examine our discussion of commutative rings to see how much of it applies to noncommutative rings (perhaps with some modification). [M2]
2. Can you find conditions on a ring such that all of its ideals are principal ideals? (Such a ring is, naturally, called a *principal ideal ring*.) [J1], [M2]
3. What kind of conditions might be imposed on an ideal R of a ring S in order that $S - R$ be a field? [M2]
4. A group G is called *simple* if its only normal subgroups are the trivial ones, G and $\{e\}$. Investigate the existence of simple groups. (Begin with the Abelian case.) [Z1]
5. A permutation on a set X is a 1–1 mapping of X into itself. If G is any permutation group on X and $x, y \in X$ we say that x is *G-equivalent* to y and write $x \sim y$ if and only if there exists a $\sigma \in G$ such that $x\sigma = y$. The equivalence classes of X are called the *orbits* of X relative to G. Investigate the relationship of these orbits to the material discussed in this chapter. [C3]

REFERENCES FOR FURTHER READING

[B4,] [C2], [C3], [F1], [H6], [J1], [K4], [M2], [Z1].

10

Complex Numbers and Quaternions

1. Complex Numbers as Residue Classes

The results of Section 9.2 suggest that we might define the set of complex numbers as the ring $C = R^*[x] - [x^2 + 1]$. Computations in C then proceed as in $R^*[x]$ except that, whenever $\overline{x^2}$ appears, we replace it by $\overline{-1}$.

Theorem 1. The ring $C = R^*[x] - [x^2 + 1]$ is a field that contains a subfield $\overline{R} \cong R^*$.

Proof. We know in general (Theorem 9.4) that the difference ring of a commutative ring is a commutative ring. It is obvious that $\overline{1}$ is a multiplicative identity for C. Furthermore, if $\overline{a + bx} \neq \overline{0}$ † then one, at least, of a and b is not zero and we observe that

$$\begin{aligned} \overline{(a + bx)}\,\overline{(a' + b'x)} &= \overline{(a + bx)(a' + b'x)} \\ &= \overline{aa' + (ab' + ba')x + bb'x^2} \\ &= \overline{(aa' - bb') + (ab' + ba')x} = \overline{1} \end{aligned}$$

if

$$a' = \frac{a}{a^2 + b^2} \qquad \text{and} \qquad b' = -\frac{b}{a^2 + b^2}$$

Finally, we define $\overline{R} = \{\overline{a} | a \in R^*\}$ and see immediately that $\overline{R} \cong R^*$ via the 1–1 correspondence, $a \in R^* \leftrightarrow \overline{a} \in \overline{R}$.

Exercises 10.1

Perform the following computations in C.

1. $\overline{2 + 3x} + \overline{-1 + x}$
2. $\overline{5 - 2x} + \overline{2 - 3x}$
3. $\overline{3 - 4x} - \overline{5 + 2x}$
4. $\overline{-5 + 2x} - \overline{-1 - 2x}$
5. $\overline{(3 - 2x)}\,\overline{(-2 + 5x)}$
6. $\overline{(-1 - x)}\,\overline{(-3 + 2x)}$

† The notation $\overline{a + bx}$ used here is, of course, the notation adopted for residue classes in Section 9.2 and is not be be confused with the notation $\overline{a + bi} = a - bi$ for the so-called *complex conjugate* of $a + bi$.

7. Find the multiplicative inverse of

(a) $\overline{2 - 3x}$ (b) $\overline{-2 + 3x}$

(c) $\overline{4 + 3x}$ (d) $\overline{7 + x}$

8. Perform the following divisions. (*Hint:* $\overline{(a + bx)} \div \overline{(c + dx)} = \overline{(a + bx)}$ $[1/\overline{(c + dx)}]$.

(a) $\overline{(2 + 3x)} \div \overline{(2 - 3x)}$ (b) $\overline{(-1 + x)} \div \overline{(2 + 5x)}$

(c) $\overline{(-1 - x)} \div \overline{(3 - 5x)}$ (d) $\overline{(-4 + 3x)} \div \overline{(-2 - 5x)}$.

2. Complex Numbers as Pairs of Real Numbers

The complex numbers may also be defined as pairs of real numbers. The procedure is very similar to that used in defining the integers and the rational numbers. Hence we shall do no more here than state the necessary definitions and leave the proofs of the various properties of the complex numbers as exercises for the student.

Definition 1. *A complex number is an ordered pair of real numbers,* (a, b), $a, b \in R^*$.

Definition 2. $(a, b) \sim (c, d)$ *if and only if* $a = c$ and $b = d$.

Here we may notice that two complex numbers are equivalent if and only if they are identical. Thus we do not need to consider equivalence classes and may write $(a, b) = (c, d)$ in place of $(a, b) \sim (c, d)$.

Definition 3. $(a, b) + (c, d) = (a + c,\ b + d)$.

(Cf. $(a + bi) + (c + di) = (a + b) + (c + d)i$.)

Definition 4. $(a,\ b) \times (c,\ d) = (ac - bd,\ ad + bc)$.

(Cf. $(a + bi)(c + di) = (ac - bd) + (ad + bc)i$.)

We leave to the student the proof of the following theorem.

Theorem 2. Let $C = \{(a, b) | a, b \in R^*\}$. Then C is a field under the operations of addition and multiplication given in Definitions 3 and 4. The identity element for addition is $(0, 0)$ and the identity element for multiplication is $(1, 0)$. Also $\bar{R} = \{(a, 0) | a \in R^*\} \subset C$ and $\bar{R} \cong R^*$ via the correspondence $a \in R^* \leftrightarrow (a, 0) \in \bar{R}$.

From now on we will use the customary notation, $a + bi$, for the complex number (a, b) or $\overline{a + bx}$.

Exercises 10.2

1. Prove Theorem 2.

2. Prove that the mapping σ defined by $(a + bi)\sigma = a - bi$ is an automorphism of C.

3. Algebraically Closed Fields

One way of picturing the "evolution" of our number system from N to I to R to R^* to C is in terms of the solution of equations. (In fact, this approach was used to motivate our definitions for the integers and for the rational numbers.) Thus natural numbers suffice to solve the equations $x + 4 = 6$ and $2x = 6$ but we need negative integers to solve $x + 6 = 4$ and non-integral rational numbers to solve $2x = 5$. Similarly we need irrational numbers to solve $x^2 - 2 = 0$ and complex, nonreal, numbers to solve $x^2 + 1 = 0$. Now none of the equations just listed are at all complicated looking; they are, in fact, linear or quadratic with integral coefficients. Thus it might seem very plausible that we would need a continuous series of extensions of our number system to solve polynomial equations of higher degree or with coefficients other than integers. For example, it is surely not evident that the equation

$$\sqrt{26}x^5 + (2 + i)x^4 - \pi x^3 + ix^2 - 5x + 1 = 0$$

has even one root that is a complex number.

Definition 5. *A field F is said to be* **algebraically closed** *if every equation $a_0x^n + a_1x^{n-1} + \cdots + a_{n-1}x + a_n = 0$ where $a_0, a_1, \ldots, a_n$ are in F has a root in F.*

Thus since $x^2 + (-2) = 0$ has coefficients in R but no root in R, R is not algebraically closed. Since $x^2 + 1 = 0$ has coefficients in R^* but no root in R^*, R^* is not algebraically closed.

Theorem 3. The complex numbers form an algebraically closed field.

All known proofs of this theorem, often called the fundamental theorem of algebra, involve nonalgebraic notions such as those found in topology or complex variable theory and hence we omit the proof here. Proofs may be found in [B4], [D2], and [V1].

4. De Moivre's Theorem

While it is not easy to show that the complex numbers form an algebraically closed field, it is easy to show that the equation $ax^n + b = 0$ ($a, b \in C$, $a \neq 0$, $n \in N$) always has a complex root. We will, in fact, describe a method for finding all of the roots of such equations. To do this we rewrite the equation $ax^n + b = 0$ as $x^n = -b/a$ and turn our attention to finding the nth roots of any complex number.

We begin by representing any complex number $a + bi$ as a point with coordinates (a, b) in the Cartesian plane as shown in the figure below. The horizontal axis is called the real axis (R) and the vertical axis the imaginary axis

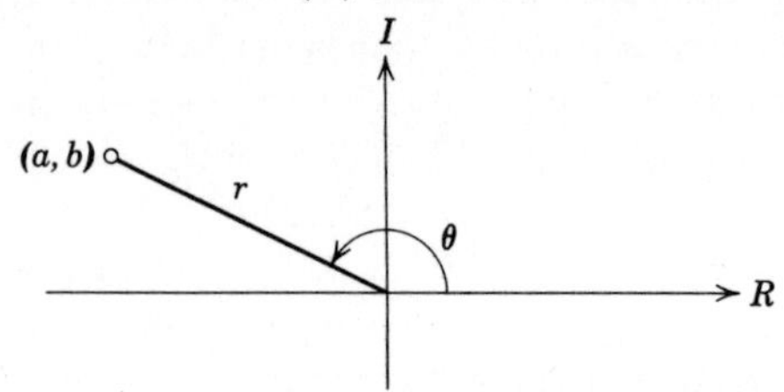

(I). With the complex number $a + bi$ is associated the real number $r = \sqrt{a^2 + b^2}$ and a positive angle θ—the angle that the line segment from (a, b) to the origin makes with the positive end of the real axis.

By simple trigonometry we have $a = r \cos \theta$ and $y = b \sin \theta$ so that

$$a + bi = r(\cos \theta + i \sin \theta)$$

The expression $r(\cos \theta + i \sin \theta)$ is called the *polar form* of the complex number $a + bi$ and is often abbreviated to $r \text{ cis } \theta$. The number r is called the *modulus* or *absolute value* of the complex number and the degree or radian measure of θ is called the *amplitude*.

Example 1. $1 + i = \sqrt{2} \text{ cis } 45°$; modulus $\sqrt{2}$, amplitude $45°$.

Example 2. $-1 - \sqrt{3}i = 2 \text{ cis } 240°$; modulus 2, amplitude $240°$.

Example 3. $4 - 3i = 5 \text{ cis } \theta$ where $\theta = 360° - \cos^{-1} 4/5$. (Recall that $0° \leq \cos^{-1} y \leq 180°$.)

Note that, for example, $\rho = 2(\sin 30° + i \cos 30°)$ is *not* in polar form but that $\rho = 2 \text{ cis } 60°$. Similarly, $\sigma = -2 \text{ cis } 60°$ is not in polar form but $\sigma = 2 \text{ cis } 240°$ is.

Theorem 4. $(\text{cis } \theta)^n = \text{cis } n\theta$ for n a non-negative integer.

Proof. The theorem is obviously true for $n = 0$ or 1. Let $S = \{n | n \in N$ and $(\text{cis } \theta)^n = \text{cis } n\theta\}$. Then $1 \in S$. Suppose $k \in S$. Then

$$\begin{aligned}
&(\text{cis } \theta)^{k+1} = (\text{cis } \theta)^k \text{ cis } \theta = \text{cis } k\theta \text{ cis } \theta \\
&= (\cos k\theta + i \sin k\theta)(\cos \theta + i \sin \theta) \\
&= (\cos k\theta \cos \theta - \sin k\theta \sin \theta) + i(\cos k\theta \sin \theta + \sin k\theta \cos \theta) \\
&= \cos (k\theta + \theta) + i \sin (k\theta + \theta) = \cos (k + 1)\theta + i \sin (k + 1)\theta \\
&= \text{cis } (k + 1)\theta
\end{aligned}$$

by use of the trigonometric identities for $\cos (\theta + \phi)$ and $\sin (\theta + \phi)$. Hence $k \in S$ implies $k + 1 \in S$ and $S = N$.

Corollary. (De Moivre's theorem) $(\text{cis } \theta)^n = \text{cis } n\theta$ for $n \in I$.

Proof. By Theorem 4 the corollary is true for $n \in N$ and for $n = 0$. Now suppose that $-n \in N$. Observe that

$$(\cos \theta + i \sin \theta)(\cos \theta - i \sin \theta) = 1$$

and that, since $\cos \theta - i \sin \theta = \cos(-\theta) + i \sin(-\theta)$, $(\text{cis } \theta)^{-1} = \text{cis}(-\theta)$. Hence if $-n \in N$, $(\text{cis } \theta)^n = [(\text{cis } \theta)^{-1}]^{-n} = [\text{cis}(-\theta)]^{-n} = \text{cis}(-n)(-\theta)$ $= \text{cis } n\theta$.

Example 4. Evaluate $(1 + i)^{-10}$.

Solution. $(1 + i)^{-10} = (\sqrt{2} \text{ cis } 45°)^{-10} = (\sqrt{2})^{-10} \text{ cis}(-450°) = (1/2^5)$ $\text{cis}(-90°) = (1/32)[\cos(-90°) + i \sin(-90°)] = -(1/32)i$.

Theorem 5. Let $\rho = r \text{ cis } \theta$ be a complex number. Then the n nth roots of ρ are given by

$$(1) \quad \sqrt[n]{r} \text{ cis} \left(\frac{\bar{\theta}}{n} + k \cdot \frac{360}{n}\right) (k = 0, 1, \ldots, n - 1)$$

where $\sqrt[n]{\ }$ is the ordinary (real) nth root of r and $\bar{\theta}$ is the degree measure of θ.

Proof. By Theorem 4

$$\left[\text{cis} \left(\frac{\bar{\theta}}{n} + k \cdot \frac{360}{n}\right)\right]^n = \text{cis}(\bar{\theta} + k \cdot 360) = \text{cis } \theta$$

for $k = 0, 1, \ldots, n - 1$. Thus the complex numbers given by (1) are nth roots of ρ. That they are all distinct follows from the fact that

$$\text{cis} \left(\frac{\bar{\theta}}{n} + i \cdot \frac{360}{n}\right) \neq \text{cis} \left(\frac{\bar{\theta}}{n} + i \cdot \frac{360}{n}\right)$$

for $i \neq j$, $i, j = 0, 1, 2, \ldots, n - 1$.

Example 5. Find the three cube roots of -1.

Solution. $-1 = \text{cis } 180°$. Thus the three cube roots of -1 are given by

$$\text{cis} \left(\frac{180°}{3} + k \cdot \frac{360°}{3}\right), \qquad k = 0, 1, 2$$

Hence they are $\text{cis } 60° = \frac{1}{2} + \frac{1}{2}\sqrt{3}i$, $\text{cis } 180° = -1$, and $\text{cis } 300° = \frac{1}{2} - \frac{1}{2}$ $\sqrt{3}i$.

Example 6. Find the two square roots of i.

Solution. $i = \text{cis } 90°$. Thus the two square roots of i are given by

$$\text{cis}\left(\frac{90°}{2} + k \cdot \frac{360°}{2}\right), \qquad k = 0, 1$$

Hence they are $\text{cis } 45° = \sqrt{2}/2 + (\sqrt{2}/2)i$ and $\text{cis } 225° = -\sqrt{2}/2 - (\sqrt{2}/2)i$.

Example 7. Find the four fourth roots of $1 + i$.

Solution. $1 + i = \sqrt{2}\ \text{cis } 45°$. Thus the four fourth roots of $1 + i$ are given by

$$\sqrt[8]{2}\ \text{cis}\left(\frac{45°}{4} + k \cdot \frac{360°}{4}\right), \qquad k = 0, 1, 2, 3$$

Hence they are $\sqrt[8]{2}$ cis 11.25°, $\sqrt[8]{2}$ cis 101.25°, $\sqrt[8]{2}$ cis 191.25°, and $\sqrt[8]{2}$ cis 281.25°. (It is possible, of course, to calculate cis 11.25°, cis 101.25° etc. in terms of radicals by using the half-angle formulas on cos 45° and sin 45° to get cos $22\frac{1}{2}°$, sin $22\frac{1}{2}°$ etc., but we prefer to leave the answer as is!)

The successful use of De Moivre's theorem depends strongly on the existence of a simple polar form of the complex number involved. Thus, for example, if we seek to find $(-2 + 3i)^3$ by the use of De Moivre's theorem, the best we can do is to write

$$-2 + 3i \cong \sqrt{13}\ \text{cis } 123°\ 40'$$

or as

$$-2 + 3i = \sqrt{13}\ \text{cis}\left[\arctan\left(-\frac{3}{2}\right) + 180°\right].$$

If we use the first (approximate) representation we obtain

$$\begin{aligned}(-2 + 3i)^3 &\cong 46.87 \text{ cis } 371° = 46.87 \text{ cis } 11° \\ &\cong 46.87\,(0.9816 + 0.1908i) \cong 46.01 + 8.943i.\end{aligned}$$

If we use the second representation it takes a considerable amount of elaborate trigonometry (see [D6]) to arrive at

$$\begin{aligned}(-2 + 3i)^3 &= 13\sqrt{13}\ \text{cis}\left[3 \arctan\left(-\frac{3}{2}\right) + 540°\right] \\ &= 13\sqrt{13}\left(\frac{46\sqrt{13}}{169} + \frac{9\sqrt{13}}{169}i\right) = 46 + 9i.\end{aligned}$$

The direct algebraic computation is definitely simpler in this case.

Exercises 10.4

1. Plot the following complex numbers: $2 + 3i$, $-1 - 2i$, -3, $4i$, $(1 + i)/(1 - i)$, $2/i$.
2. Find the amplitude and absolute value of each of the following complex numbers:
   $4(\cos 40° + i \sin 40°)$, $-2(\cos 15° + i \sin 15°)$,
   $3(\sin 15° + i \cos 15°)$, $3(\cos 40° - i \sin 40°)$,
   $3(-\cos 30° + i \sin 30°)$
3. Find the polar form of AB, A/B, A^2/B, and $1/A$ if $A = 3 \text{ cis } 60°$ and $B = 4 \text{ cis } 40°$.
4. Prove that the absolute value of the product of two complex numbers is the product of their absolute values, and that the amplitude of the product is the sum of their amplitudes.
5. What is the geometrical locus of the set of complex numbers of (a) fixed absolute value, (b) of fixed amplitude?
6. Write the numbers $A = 1 - \sqrt{3}i$, $B = -1 - \sqrt{3}i$, and $C = -1 + \sqrt{3}i$ in polar form. Find the amplitude and absolute value of A^2 and of B^2/AC.
7. Use De Moivre's theorem to calculate:
   (a) $(1 + i)^6$ (b) $(1 - \sqrt{3}i)^5$
   (c) $(\sqrt{3} - i)^7$ (d) $(1 - i)^{10}$
8. Use De Moivre's theorem and the binomial theorem to find expressions for $\cos 3\theta$ and $\sin 3\theta$ in terms of $\cos \theta$ and $\sin \theta$
9. Find the three cube roots of 8.
10. Find the three cube roots of $8i$.
11. Find the four fourth roots of 1.
12. Find the two square roots of $(1 + i)/\sqrt{2}$. (Leave your answer in trigonometric form.)
13. Find the three cube roots of $1 - \sqrt{3}i$. (Leave your answer in trigonometric form.)
14. Prove that there exists a complex number R such that the n nth roots of 1 are given by $R, R^2 \ldots, R^n$.
15. A complex number z is called a *primitive* nth root of 1 if $z^n = 1$ but $z^m \neq 1$ when $0 < m < n$.
    (a) Find the primitive sixth roots of 1.
    *(b) Prove that a necessary and sufficient condition that R^k (problem 14) be a primitive nth root of 1 is that $(k, n) = 1$.

*16. Use De Moivre's theorem to find
   (a) $(1 - 2i)^4$ (b) $(3 + 2i)^3$
   without the use of trigonometric tables.

5. Quaternions

A system which has all of the properties of a field except that of commutativity of multiplication is called a *skew-field*. The first example of a skew-field was given by Hamilton in 1843, and it served as a stimulus to the construction later of many other abstract systems. The system that Hamilton developed is called the algebra of *quaternions* over the field R^*.

More generally, however, we may consider a *quaternion algebra* over any field F as the set

$$Q = \{a_0 + a_1 i + a_2 j + a_3 k | a_0, a_1, a_2, a_3 \in F\}$$

where addition in Q is defined by

$$(a_0 + a_1 i + a_2 j + a_3 k) + (b_0 + b_1 i + b_2 j + b_3 k)$$
$$= (a_0 + b_0) + (a_1 + b_1)i + (a_2 + b_2 j) + (a_3 + b_3)k$$

and equality by

$$a_0 + a_1 i + a_2 j + a_3 k = b_0 + b_1 i + b_2 j + b_3 k$$

if and only if $a_0 = b_0$, $a_1 = b_1$, $a_2 = b_2$, and $a_3 = b_3$. The additive identity for Q is $0 + 0i + 0j + 0k$ and the additive inverse of $a_0 + a_1 i + a_2 j + a_3 k$ is $-a_0 + (-a_1)i + (-a_2)j + (-a_3)k = -a_0 - a_1 i - a_2 j - a_3 k$. The *scalar product* $a(a_0 + a_1 i + a_2 j + a_3 k) = aa_0 + aa_1 i + aa_2 j + aa_3 k$ for any $a \in F$.

Multiplication of elements in Q is performed just as for polynomials except that we replace i^2, j^2, and k^2 by -1 whenever the occur and ij by k, ji by $-k$, ik by $i(ij) = i^2 j = -j$, ki by $(ij)i = i(ji) = -i(ij) = -i^2 j = j$ etc. (See multiplication table for the "units" below.)

	i	j	k
i	-1	k	$-j$
j	$-k$	-1	i
k	j	$-i$	-1

Example. $(1 + i - k)(2 + 4j - 3k) = 2 + 4j - 3k + 2i + 4ij - 3ik - 2k - 4kj + 3k^2 = 2 + 4j - 3k + 2i + 4k + 3j - 2k + 4i - 3 = -1 + 6i + 7j - k$.

The student should check that if $\alpha = a_0 + a_1 i + a_2 j + a_3 k$ where $d = a_0^2 + a_1^2 + a_2^2 + a_3^2 \neq 0$, then the multiplicative inverse of α is

$$\alpha^{-1} = a_0' - a_1' i - a_2' j - a_3' k$$

where

$$a_1' = a_i / d \qquad (i = 0, 1, 2, 3)$$

If, in addition to not requiring commutativity of multiplication, we do not require multiplicative inverses for all elements, it is possible to construct in a similar fashion, many other such "linear algebras" over any given field. For details, see the references given for this chapter.

Exercises 10.5

1. Perform the indicated operations:
(a) $(2 + 3i - 4j - 2k) + (-3 + i - 5j + 6k)$
(b) $(-3 + 2i - 5j + k) + (-1 - 3i + 2j - k)$
(c) $(1 - i - 2j + k)(2 + i - j + k)$
(d) $(-1 + i - j + 2k)(3 - i + 2j - k)$

2. Find the multiplicative inverses of the following quaternions:
(a) $2 + 3i - 4j - 2k$ (b) $-3 + 2i - 5j + k$
(c) $1 + i + j + k$ (d) $-1 - i + j - k$

3. Perform the following divisions:
(a) $(1 + i - k) \div (2 + 4j - k)$
(b) $(1 - i + j - k) \div (1 + i - j + 2k)$
(c) $(i - j + k) \div (1 - j + k)$
(d) $(1 - j + k) \div (i + j + k)$

4. The *norm*, $N(\alpha)$, of the quaternion $\alpha = a_0 + a_1i + a_2j + a_3k$ is defined to be $a_0^2 + a_1^2 + a_2^2 + a_3^2$. Prove that the norm of the product of two quaternions is the product of their norms.

5. The *conjugate* of $\alpha = a_0 + a_1i + a_2j + a_3k$ is defined to be $\bar{\alpha} = a_0 - a_1i - a_2j - a_3k$. Prove that if α and β are quaternions, then $\overline{\alpha\beta} = \bar{\beta}\bar{\alpha}$ and $\bar{\alpha}\alpha = \alpha\bar{\alpha} = N(\alpha)$.

PROBLEMS FOR FURTHER INVESTIGATION

1. Investigate the solution of polynomial equations $a_0x^n + a_1x^{n-1} + \ldots + a_n = 0$ when $a_0, a_1, \ldots, a_n$ are quaternions. Begin with linear equations, quadratic equations, and equations of the form $x^n + a = 0$. In particular, see if you can find a De Moivre type theorem for roots of quaternions. [B4], [B5], [N2]

2. Investigate the possibility of developing an algebra with units $1, i_1, i_2, \ldots, i_8$ along the lines of the quaternions. (Be prepared to sacrifice the associative law of multiplication!) [A2], [D3], [D5]

3. Let $Q = \{a_0 + a_1i + a_2j + a_3k \mid a_0, a_1, a_2, a_3 \in I\}$. Develop a theory of divisibility in Q along the lines of Section 5.3.

REFERENCES FOR FURTHER READING

[A1], [A2], [B4], [D3], [D5], [N2].

11

Vector Spaces

1. Definition

We have already seen many instances of the tendency of the mathematician to abstract certain key properties of familiar systems and to consider more general systems satisfying these properties. Thus integral domains may be thought of as a generalization of the integers, fields as generalizations of the rational numbers etc. Here our generalization arises from the ordinary notion of a vector in two or three dimensions. If we consider vectors as being directed line segments originating from a fixed point we may define their addition by the "parallelogram law" as illustrated below for the plane: the vector α added to the vector β gives the vector γ. A further concept is that of a

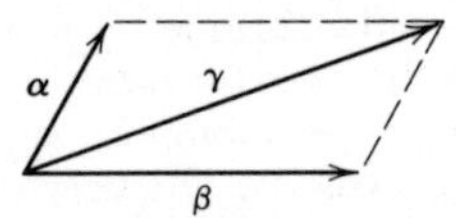

a scalar product, $c\alpha$, of a vector α by a real number c as illustrated below.

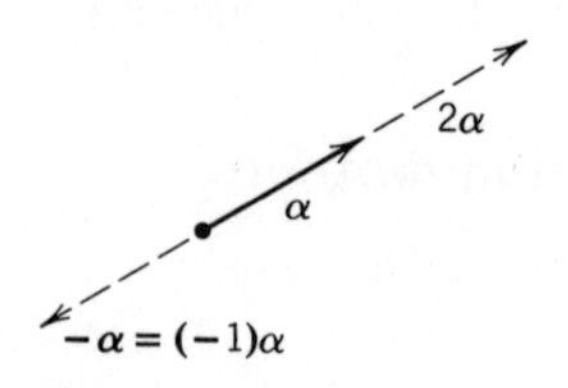

It is easy to verify that ordinary vectors in two or three dimensions satisfy the requirements for a vector space in the sense of the following definition.

Definition 1. *A **vector** (or **linear**) **space** V over a field F is a set of elements, called **vectors,** such that $\alpha \in V$ and $\beta \in V$ define a (unique) $\alpha + \beta \in V$ and for any $\alpha \in V$ and **scalar** $c \in F$ there is defined a **scalar** product $c\alpha \in V$ such that*
(1) V is an Abelian group under addition
(2) $c(\alpha + \beta) = c\alpha + c\beta$, $(c + c')\alpha = c\alpha + c'\alpha$

(3) $(cc')\alpha = c(c'\alpha)$, $1\alpha = \alpha$
for all $c, c' \in F$, $\alpha, \beta \in V$, *and* 1 *the multiplicative identity of* F.

To avoid confusion we will designate the identity for addition in F by 0 and the identity for addition in V by $\bar{0}$.

The student should check that the following are examples of vector spaces.

Example 1. For any fixed $n \in N$, $V_n(F) = \{(a_1, \ldots, a_n) | a_i \in F$ for $i = 1, \ldots, n\}$ where $(a_1, \ldots, a_n) + (b_1, \ldots, b_n) = (a_1 + b_1, \ldots, a_n + b_n)$ and $c(a_1, \ldots, a_n) = (ca_1, \ldots, ca_n)$ is a vector space over F. (This space is called the *vector space of n-tuples*.)

Example 2. $C = \{a + bi | a, b \in R^*\}$ is a vector space over R^*. That is, the complex numbers may be considered as a vector space over the real numbers.

Note that when we write, for example, the scalar product of $2(2 + 0i)$ we are properly distinguishing between the scalar (real number), 2, and the vector (complex number), $2 + 0i$. We know, of course, that the set of complex numbers of the form $a + 0i$ is isomorphic to R^* and sometimes abbreviate $2(2 + 0i)$ as $2 \cdot 2$. In general, when a vector space V over F contains a subset $F^* \cong F$ we often use the same symbol for the elements of F^* as for those of F.

Example 3. $\{a_0 + a_1x + \ldots + a_nx^n | n \in N, a_i \in R, i = 1, \ldots, n\}$ is a vector space over R. That is, the set of all polynomials with rational coefficients may be considered as a vector space over R.

Definition 2. *Let V be a vector space over a field F. Then S is said to be a* ***subspace*** *of V if and only if $S \subseteq V$ and S is itself a vector space over F.*

Example 4. $S = \{(a_1, a_2, 0) | a_1, a_2 \in F\}$ is a subspace of $V_3(F)$ over F.

Example 5.

$$S = \left\{ \sum_{i=1}^{m} b_i x^{2i} | m \in N, b_i \in R, i = 1, \ldots, m \right\}$$

is a subspace of the vector space of all polynomials with rational coefficients.

Example 6. Any vector space is a subspace of itself.

Theorem 1. Let V be a vector space over a field F. Then
(1) $0\alpha = \bar{0}$ for all $\alpha \in V$;
(2) $c\bar{0} = \bar{0}$ for all $c \in F$;
(3) The additive inverse, $-\alpha$, of α is equal to $(-1)\alpha$.

Proof. (1) We have $(c + 0)\alpha = c\alpha = c\alpha + \bar{0} = c\alpha + 0\alpha$. By the cancellation property for groups we have $\bar{0} = 0\alpha$.

(2) $c(\alpha + \bar{0}) = c\alpha = c\alpha + \bar{0} = c\alpha + c\bar{0}$ so that $\bar{0} = c\bar{0}$.

(3) $\alpha + (-1)\alpha = 1\alpha + (-1)\alpha = [1 + (-1)]\alpha = 0\alpha = \bar{0}$ so that $(-1)\alpha = -\alpha$.

Exercises 11.1

1. Prove that, in any vector space, $c\alpha = \bar{0}$ implies $c = 0$ or $\alpha = \bar{0}$.
2. Let $\alpha = (1, 0, 1)$, $\beta = (2, -1, 1)$, and $\gamma = (3, -1, 0)$. Compute
   (a) $2(\alpha - 3\beta) + 4\gamma$
   (b) $2\alpha - 3\beta + 4(\alpha - \beta + \gamma)$
3. Which of the following subsets of $V_n(R)$ are subspaces of $V_n(R)$?
   (a) $\{(x_1, x_2, \ldots, x_n) | x_2 = 0\}$
   (b) $\{(x_1, x_2, \ldots, x_n) | x_2 \in I\}$
   (c) $\{(x_1, x_2, \ldots, x_n) | x_1 = 0 \text{ or } x_2 = 0\}$
   (d) $\{(x_1, x_2, \ldots, x_n) | x_2 + 2x_3 = 0\}$
   (e) $\{(x_1, x_2, \ldots, x_n) | x_1 \neq 0\}$
   (f) $\{(x_1, x_2, \ldots, x_n) | x_1 + x_2 + x_3 = 0\}$
4. A function f with *domain* S is a mapping of S into T: $x \in S \to f(x) \in T$ (Section 7.5). Suppose $T = R^*$ and define $f + g$ as the mapping $x \to f(x) + g(x)$ and cf for $c \in R^*$ as the mapping $x \to cf(x)$. Which of the following sets of functions defined on $S = \{x | x \in R^*, -1 \leq x \leq 1\}$ are vector spaces?
   (a) $\{f | f(x) = a_0 + a_1x + a_2x^2, a_1, a_2, a_3 \in R^*\}$
   (b) $\{f | f \text{ is continuous on } -1 \leq x \leq 1\}$
   (c) $\{f | f(x) \geq 0\}$
   (d) $\{f | f(2) = 0\}$
   (e) $\{f | 2f(0) = f(1)\}$
   (f) $\{f | f(x) = f(-x)\}$
5. If S and T are two subspaces of a vector space V over F, prove that $S \cap T$ is a subspace of V.

*6. Prove commutativity of addition of vectors from the other properties of a vector space: (*Hint:* Expand $(1 + 1)(\alpha + \beta)$ in two ways.)

2. The Basis of a Vector Space

Definition 3. *Let S be a set of vectors of a vector space V over F. Then S is said to **span** or **generate** V if and only if $\alpha \in V$ implies that there exist $\xi_1, \ldots, \xi_n \in S$ and $c_1, \ldots, c_n \in F$ such that*

$$\alpha = \sum_{i=1}^{n} c_i \xi_i$$

Example 1. In $V_3(F)$ the vectors $\xi_1 = (1, 0, 0)$, $\xi_2 = (0, 1, 0)$, and $\xi_3 = (0, 0, 1)$ span $V_3(F)$ since $(a_1, a_2, a_3) = a_1\xi_1 + a_2\xi_2 + a_3\xi_3$. Also ξ_1, ξ_2, ξ_3, and $\xi_4 = (1, 1, 1)$ span $V_3(F)$ since $(a_1, a_2, a_3) = a_1\xi_1 + a_2\xi_2 + a_3\xi_3 + 0\xi_4$, but ξ_1 and ξ_2 alone do not span $V_3(F)$.

Example 2. The set $S = \{x^n | n = 0 \text{ or } n \in N\} \cup \{0\}$ is a spanning set for the vector space over R of all polynomials with rational coefficients.

Definition 4. *A finite set $\{\alpha_1, \ldots, \alpha_n\}$ of vectors of a vector space V over F is said to be* **linearly independent** *if, for $c_1, \ldots, c_n \in F$,*

$$\sum_{i=1}^{n} c_i\alpha_i = \bar{0}$$

implies $c_1 = \ldots = c_n = 0$. An infinite set S, of vectors of V is said to be linearly independent if every finite subset of S is linearly independent. A set that is not linearly independent is said to be **linearly dependent**.

Example 3. In Example 1 the vectors ξ_1, ξ_2, and ξ_3 are clearly linearly independent since $c_1\xi_1 + c_2\xi_2 + c_3\xi_3 = (c_1, c_2, c_3) = (0, 0, 0)$ implies $c_1 = c_2 = c_3 = 0$. But ξ_1, ξ_2, ξ_3, and ξ_4 are linearly dependent since $1\xi_1 + 1\xi_2 + 1\xi_3 + (-1)\xi_4 = (0, 0, 0)$.

Example 4. In Example 2 the set S is a linearly independent set but the set $S' = S \cup \{1 + x\}$ is not since $1 \cdot 1 + 1\,x + (-1)(1 + x) = \bar{0}$. (Note the dual use of "1" in "$1 \cdot 1$" as a symbol for both the rational number 1 and the polynomial $1 \cdot x^0$.)

Example 5. Suppose that the vectors ξ_1, ξ_2, and ξ_3 are linearly independent in $V_3(R)$. Then the vectors $\xi_1 + \xi_2$, $\xi_1 + \xi_3$, and $\xi_2 + \xi_3$ are also linearly independent over R.

Proof. We have given that $c_1\xi_1 + c_2\xi_2 + c_3\xi_3 = \bar{0}$ implies that $c_1 = c_2 = c_3 = 0$. Suppose that for some $a_1, a_2, a_3 \in R$, $a_1(\xi_1 + \xi_2) + a_2(\xi_1 + \xi_3) + a_3(\xi_2 + \xi_3) = \bar{0}$. Then $(a_1 + a_2)\xi_1 + (a_1 + a_3)\xi_2 + (a_2 + a_3)\xi_3 = \bar{0}$. Thus $a_1 + a_2 = a_1 + a_3 = a_2 + a_3 = 0$ and it follows that $a_1 = a_2 = a_3 = 0$ so that $\xi_1 + \xi_2$, $\xi_1 + \xi_3$, and $\xi_2 + \xi_3$ are linearly independent over R.

It is convenient to make the arbitrary agreement that the null set is a linearly independent subset of any vector space V.

Definition 5. *A subset S of a vector space V is said to be a* **basis** *of V if* (1) *S spans V and* (2) *S is a linearly independent set.*

Roughly speaking, a basis of a vector space V is a minimal set of vectors needed to generate V.

Example 6. In $V_n(F)$ the *unit* vectors $\xi_1, \ldots, \xi_n$ where ξ_i is the vector $(0, \ldots, 0, 1, 0, \ldots, 0)$ with 1 in the ith place and zeros elsewhere clearly form a basis.

Example 7. $S = \{x^n | n = 0 \text{ or } n \in N\} \cup \{0\}$ is a basis for the vector space over R of all polynomials with rational coefficients.

Example 8. If the vectors ξ_1, ξ_2, and ξ_3 form a basis of V over R so do $\xi_1 + \xi_2$, $\xi_1 + \xi_3$, and $\xi_2 + \xi_3$.

Proof. We have already shown (Example 5) that $\xi_1 + \xi_2$, $\xi_1 + \xi_3$, and $\xi_2 + \xi_3$ are linearly independent if ξ_1, ξ_2, and ξ_3 are. Now we observe that if $\alpha = c_1\xi_1 + c_2\xi_2 + c_3\xi_3$, then

$$\alpha = (c_1/2)(\xi_1 + \xi_2) + (c_2/2)(\xi_1 + \xi_3) + (c_3/2)(\xi_2 + \xi_3)$$

Hence $\xi_1 + \xi_2$, $\xi_1 + \xi_3$, and $\xi_2 + \xi_3$ also span V and hence form a basis of V.

The student should avoid two common errors. One is to assume that every vector space is a space of n-tuples† and the other is to assume that, in a space of n-tuples, the basis is always one of unit vectors. Thus the vector space of all polynomials with rational coefficients is not a space of n-tuples (although it can, as we shall see in the next chapter, be regarded as a space with elements of the form $(a_1, a_2, \ldots)$ with an infinite number of entries). Likewise in $V_n(F)$ there are other bases besides the set of unit vectors. Thus, for example, the vectors $(1, 1, 0)$, $(1, 0, 1)$, and $(0, 1, 1)$ are a basis of $V_3(F)$ since if $\xi_1 = (1, 0, 0)$, $\xi_2 = (0, 1, 0)$, and $\xi_3 = (0, 0, 1)$ we have $\xi_1 + \xi_2 = (1, 1, 0)$, $\xi_1 + \xi_3 = (1, 0, 1)$, and $\xi_2 + \xi_3 = (0, 1, 1)$. (See Example 8.)

Exercises 11.2

1. Prove that the vectors (a_1, a_2) and (b_1, b_2) in $V_2(F)$ are linearly independent if and only if $a_1b_2 - a_2b_1 \neq 0$.
2. In $V_3(R)$, examine each of the following sets of vectors for linear dependence.
   (a) $(1, 3, 2)$ and $(1, 4, 1)$
   (b) $(6, 4, 12)$ and $(3, 2, 6)$
   (c) $(1, 2, 3)$ and $(1, 4, 9)$
   (d) $(1, 2, 3)$, $(1, 0, 1)$, and $(0, 1, 0)$
   (e) $(1, 2, 0)$, $(0, 3, 1)$, and $(-1, 0, 1)$
   (f) $(-1, 2, 1)$, $(3, 0, -1)$, and $(-5, 4, 3)$
3. Prove that α and $\beta \neq 0$ are linearly dependent in a vector space V over F if and only if $\alpha = c\beta$ for $c \in F$.
4. Do the results of Example 5 hold over any field F?
5. How many elements are there in each subspace spanned by three linearly independent elements of $V_5(J_3)$. Generalize your result.

† We shall show later that every vector space spanned by a *finite* number of vectors is *isomorphic* to a space of n-tuples.

6. If $c_1\alpha + c_2\beta + c_3\gamma = \bar{0}$ where $c_1c_3 \neq 0$, show that α and β generate the same subspace as do β and γ.
7. Exhibit three different bases for $V_3(R)$.

3. The Dimension of a Vector Space

Our next sequence of theorems will lead us to the concept of the dimension of a vector space spanned by a finite number of elements.

Theorem 2. If $\{\xi_1, \ldots, \xi_n\}$ is a basis of V over F, then so are

(1) $\{\xi_1, \ldots, \xi_{i-1}, k\xi_i, \xi_{i+1}, \ldots, \xi_n\}$ $(k \neq 0 \in F)$

and

(2) $\{\xi_1, \ldots, \xi_{i-1}, \xi_i - k\xi_j, \xi_{i+1}, \ldots, \xi_n\}$ $(k \in F, i \neq j)$.

Proof. (1) If $c_1\xi_1 + \ldots + c_i(k\xi_i) + \ldots + c_n\xi_n = \bar{0}$ with not all of $c_1, \ldots, c_n$ equal to zero, then we have $c_1\xi_1 + \ldots + (c_ik)\xi_i + \ldots + c_n\xi_n = \bar{0}$ with not all of $c_1, \ldots, c_{i-1}, c_ik, c_{i+1}, \ldots, c_n$ equal to zero so that $\xi_1, \ldots, \xi_n$ would be linearly dependent. Hence $\xi_1, \ldots, \xi_{i-1}, k\xi_i, \xi_{i+1}, \ldots, \xi_n$ are linearly independent.

Next, if

$$\xi = \sum_{i=1}^{n} c_i\xi_i,$$

then

$$\xi = c_1\xi_1 + \cdots + c_{i-1}\xi_{i-1} + \frac{c_i}{k}(k\xi_i) + c_{i+1}\xi_{i+1} + \cdots + c_n\xi_n$$

and hence $\xi_1, \ldots, \xi_{i-1}, k\xi_i, \xi_{i+1}, \ldots, \xi_n$ span V. Thus $\{\xi_1, \ldots, \xi_{i-1}$ $k\xi_i, \xi_{i+1}, \ldots, \xi_n\}$ is a basis of V.

(2) If $c_1\xi_1 + \ldots + c_{i-1}\xi_{i-1} + c_i(\xi_i - k\xi_j) + c_{i+1}\xi_{i+1} + \ldots + c_n\xi_n = \bar{0}$, we have $c_1\xi_1 + \ldots + c_{i-1}\xi_{i-1} + c_i\xi_i + c_{i+1}\xi_{i+1} + \ldots + c_{j-1}\xi_{j-1} + (c_j - kc_i)\xi_j + c_{j+1}\xi_{j+1} + \ldots + c_n\xi_n = \bar{0}$ and, hence, since $\xi_1, \ldots, \xi_n$ are linearly independent we have $c_1 = \ldots = c_i \ldots = c_{j-1} = c_j - kc_i = c_{j+1} = \ldots = c_n = 0$. Thus $c_j = 0$ also and $\xi_1, \ldots, \xi_{i-1}, \xi_i - k\xi_j, \xi_{i+1}, \ldots, \xi_n$ are linearly independent.

Next, if

$$\xi = \sum_{i=1}^{n} c_i\xi_i,$$

then $\xi = c_1\xi_1 + \ldots + c_i(\xi_i - k\xi_j) + \ldots + (c_j + c_ik)\xi_j + \ldots + c_n\xi_n$ and hence $\xi_1, \ldots, \xi_i - k\xi_j, \ldots, \xi_n$ span V. Thus $\{\xi_1, \ldots, \xi_{i-1}, \xi_i - k\xi_j, \xi_{i+1}, \ldots, \xi_n\}$ is a basis of V.

Example. If $\{\xi_1, \xi_2, \xi_3, \xi_4\}$ is a basis of $V_4(R)$, then so are $\{\xi_1, \xi_2, 3\xi_3, \xi_4\}$ and $\{\xi_1, \xi_2 - 2\xi_3, \xi_3, \xi_4\}$.

Theorem 3. A set of nonzero vectors, $\{\xi_1, \ldots, \xi_m\}$, of V is linearly dependent if and only if some one of the ξ_k is a linear combination of the preceding vectors $\xi_1, \ldots, \xi_{k-1}$.

Proof. (1) If $\xi_k = c_1\xi_1 + \ldots + c_{k-1}\xi_{k-1}$ we have $c_1\xi_1 + \ldots + c_{k-1}\xi_{k-1} + (-1)\xi_k + 0\xi_{k+1} + \ldots + 0\xi_m = \bar{0}$ with $(-1) \neq 0$ and hence (since $\xi_k \neq \bar{0}$) $\xi_1, \ldots, \xi_m$ are linearly dependent.

(2) Suppose

$$\sum_{i=1}^{m} d_i\xi_i = \bar{0}$$

with $d_k \neq 0, d_{k+1} = \ldots = d_m = 0$ (i.e. d_k is the last nonzero d). Then if $k \neq 1$, we solve for ξ_k and are done. But if $k = 1$, $d_1\xi_1 = \bar{0}, d_1 \neq 0$ and $\xi_1 = \bar{0}$, a contradiction.

Theorem 4. A set $S = \{\xi_1, \ldots, \xi_m\}$ of vectors of V is linearly dependent if and only if it contains a proper subset R such that S and R generate the same subspace of V.

Proof. (1) If S is a linearly dependent set we form R by deleting from S all zero vectors and those which are (by Theorem 3) linear combinations of the vectors preceding it in the list. By Theorem 3 at least one vector will be deleted so that $R \subset S$.

(2) Suppose now that S contains a proper subset, R, which generates the same subspace of V as S. By renumbering our vectors if necessary we may assume $R = \{\xi_1, \ldots, \xi_k\}$ where $k < m$. Since S and R generate the same subspace, $\xi_{k+1} = c_1\xi_1 + \ldots + c_k\xi_k$ and hence $c_1\xi_1 + \ldots + c_k\xi_k + (-1)\xi_{k+1} + 0\xi_{k+2} + \ldots + 0\xi_m = \bar{0}$ so that $\{\xi_1, \ldots, \xi_m\}$ is a dependent set of vectors.

Theorem 5. If $\alpha_1, \ldots, \alpha_n$ span V and $\xi_1, \ldots, \xi_r$ are linearly independent vectors of V, then $n \geq r$.

Proof. Since $A_0 = \{\alpha_1, \ldots, \alpha_n\}$ spans V it follows by Theorem 4 that $B_0 = \{\xi_1, \alpha_1, \ldots, \alpha_n\}$ spans V and is a linearly dependent set. By Theorem 3 some one of $\xi_1, \alpha_1, \ldots, \alpha_n$ is a linear combination of its predecessors in the list. Certainly this is not ξ_1. It is, say, α_i. But then $A_1 = \{\xi_1, \alpha_1, \ldots, \alpha_{i-1}, \alpha_{i+1}, \ldots, \alpha_n\}$ spans V.

Now repeat: $B_1 = \{\xi_2, \xi_1, \alpha_1, \ldots, \alpha_{i-1}, \alpha_{i+1}, \ldots, \alpha_n\}$ spans V and is a linearly dependent set. Because $\xi_1, \ldots, \xi_r$ are linearly independent, ξ_1 is not a multiple of ξ_2 and we again may delete some $\alpha_j \neq \alpha_i$.

Now if $n < r$ we would arrive at $A_{n-1} = \{\xi_{n-1}, \ldots, \xi_1, \alpha_k\}$ spanning V and then have $B_{n-1} = \{\xi_n, \xi_{n-1}, \ldots, \xi_1, \alpha_k\}$ a linearly dependent set spanning V. As before delete an α (which now must be α_k) and arrive at $A_n = \{\xi_n, \xi_{n-1}, \ldots, \xi_1\}$ spanning V so that ξ_{n+1} is a linear combination of $\xi_1, \ldots, \xi_n$, a contradiction. Hence $n \geq r$.

Theorem 6. If $A = \{\alpha_1, \ldots, \alpha_n\}$ and $B = \{\beta_1, \ldots, \beta_m\}$ are two bases of a vector space V, then $n = m$.

Proof. For A and B span V and are linearly independent; hence we may apply Theorem 5 to get $n \geq m$ and $m \geq n$ so that $m = n$.

Definition 6. *A vector space V over F is said to be* ***finite dimensional*** *if there exists a finite set of vectors of V that span V. A vector space that is not finite dimensional is said to have* ***infinite dimension.***

We have seen (Theorem 6) that any two bases of a finite dimensional vector space have the same number of elements. This fact leads to the following definition.

Definition 7. *The* ***dimension*** *of a finite dimensional vector space V over F is the number of elements in a basis of V.*

Since a basis of $V_n(F)$ is the set of the n unit vectors of $V_n(F)$ we have the following basic result.

Theorem 7. $V_n(F)$ is a vector space of dimension n.

Thus $V_2(F)$ and $V_3(F)$ have dimension 2 and 3 respectively—as we would expect from the geometric motivation previously considered.

Definition 8. *Two vector spaces V and V' over the same field F are said to be* ***isomorphic*** *if there is a 1–1 correspondence between the elements of V and V' that is preserved under addition and scalar multiplication. That is, if $\alpha \in V \leftrightarrow \alpha' \in V'$ and $\beta \in V \leftrightarrow \beta' \in V'$, then $\alpha + \beta \in V \leftrightarrow \alpha' + \beta' \in V'$ and $c\alpha \in V \leftrightarrow c\alpha' \in V'$ for every $c \in F$.*

Theorem 8. A vector space V of dimension n over F is isomorphic to $V_n(F)$.

Proof. Let $\{\alpha_1, \ldots, \alpha_n\}$ be a basis for V and $\{\xi_1, \ldots, \xi_n\}$ be a basis for $V_n(F)$. Then if

$$\beta = \sum_{i=1}^{n} c_i\alpha_i \in V$$

we let

$$\beta \leftrightarrow \beta' = \sum_{i=1}^{n} c_i \xi_i \in V_n(F),$$

Obviously this correspondence is preserved under addition and scalar multiplication.

Corollary. Any two vector spaces of dimension n are isomorphic.

The proof of Theorem 8 is deceptively simple—so simple that the student may feel that it should have been presented in the very beginning, and thus have eliminated any need for considering the abstract finite dimensional case. In order even to state the theorem, however, we need the concept of dimension. But to talk about dimension we need to know that any two bases of a vector space have the same number of elements. In turn, this demands other information and, in fact, it would turn out that all of our preliminary work was indeed necessary. In addition, in spite of the existence of Theorem 8, it is still often much more useful to consider a finite dimensional vector space abstractly rather than as a set of n-tuples.

Exercises 11.3

1. Find a basis for $V_3(R)$ which includes the vectors $(1, -1, 1)$ and $(2, 1, 1)$.
2. Prove that no basis of $V_3(R)$ can include both of the vectors $(1, -1, 1)$ and $(2, -2, 2)$.
3. Prove that the vectors $(1, 2, 3, 4)$, $(2, 3, 1, 5)$, $(3, 1, 2, 7)$, and $(1, 1, -2, 1)$ in $V_4(R)$ are linearly dependent. Which of these four vectors is not in the subspace spanned by the other three?
4. Prove that for n vectors $\xi_1, \ldots, \xi_n$ of an n-dimensional vector space V to be a basis, it is sufficient that they span V or that they be linearly independent.
5. If two subspaces S and T of a vector space V over F have the same dimension, prove that $S \subseteq T$ implies $S = T$.

4. Fields as Vector Spaces

It is sometimes useful to consider certain fields as vector spaces over subfields. We did this, for example, when we considered $C = \{a + bi \mid a, b \in R^*\}$ as a vector space (of dimension 2) over R^*. In so doing we noted that, for example, $2 + 0i \in C$ is really not the same as $2 \in R^*$ but rather that the subset, $\bar{R} = \{a + 0i \mid a \in R^*\}$, of C is isomorphic to R^* via the correspondence $a + 0i \in \bar{R} \leftrightarrow a \in R^*$. In practice, however, we need not always maintain these distinctions and will not do so here.

Suppose now that we consider the vector space $V = \{a + b\sqrt{2} | a, b \in R\}$ over R. It has dimension two over R (why?) and hence any three vectors of V are linearly dependent. Now let $\alpha = a + b\sqrt{2} \in V$. Although we have not defined products of vectors we know that V is indeed a field as well as a vector space; thus we may consider $\alpha^2 = \alpha\alpha$ and observe that $\alpha^2 \in V$. Now consider the vectors 1, α, and α^2 of V. Since V has dimension two we know that 1, α, and α^2 are linearly dependent in R. Hence there exist rational numbers c_1, c_2, and c_3, not all zero, such that $c_1 + c_2\alpha + c_3\alpha^2 = 0$. That is, we have shown that every number of the form $a + b\sqrt{2}$, $a, b \in R$, satisfies a quadratic equation with rational coefficients.

Now it is not too difficult to find by elementary algebra the precise quadratic equation satisfied by $a + b\sqrt{2}$. The computation, however, is fairly lengthy compared with the simple existence type argument just presented. The corresponding direct method would be very much more difficult in our next example where we wish to show that every number of the form $a + b\sqrt{2} + c\sqrt{3}$, $a, b, c \in R$, satisfies a quartic equation with coefficients in R. To do this we consider $V = \{a + b\sqrt{2} + c\sqrt{3} | a, b, c \in R\}$ over R. Now we cannot use the same scheme as we did for $a + b\sqrt{2}$ since if $\alpha = a + b\sqrt{2} + c\sqrt{3}\,(bc \neq 0 \in R)$, $bc\sqrt{2}\sqrt{3} = bc\sqrt{6}$ is a term of α^2 and $\alpha^2 \notin V$. However, since $\sqrt{2}\sqrt{3} = \sqrt{6}$, $\sqrt{2}\sqrt{6} = \sqrt{12} = 2\sqrt{3}$, and $\sqrt{3}\sqrt{6} = \sqrt{18} = 3\sqrt{3}$ we see that, for any positive integer n, α^n can be written as $a' + b'\sqrt{2} + c'\sqrt{3} + d'\sqrt{6}$ with a', b', c', and $d' \in R$. Hence we consider, in place of V, $V' = \{a + b\sqrt{2} + c\sqrt{3} + d\sqrt{6} | a, b, c, d \in R\}$ over R. Then V' is a vector space of dimension at most four† over R and since $\alpha = a + b\sqrt{2} + c\sqrt{3} \in V'$ and also 1, α^2, α^3, $\alpha^4 \in V'$ we conclude that 1, α, α^2, α^3, α^4 are linearly dependent over R. Thus there exist c_1, c_2, c_3, c_4, and c_5, not all zero, in R such that $c_1 + c_2\alpha + c_3\alpha^2 + c_4\alpha^3 + c_5\alpha^4 = 0$.

As a final example, which should reveal the full generality of the process, consider the problem of showing that every number of the form $\alpha = a + b\sqrt[3]{2} + c\sqrt{3}$, $a, b, c \in R$, satisfies some polynomial equation with coefficients in R. When we consider $\alpha^2, \alpha^3, \ldots$ we obtain many other irrational numbers that are linearly independent over R when considered with 1, $\sqrt[3]{2}$, and $\sqrt{3}$. Thus we obtain $\sqrt[3]{2}\sqrt{3} = \sqrt[6]{4}\sqrt[6]{27} = \sqrt[6]{108}$, $\sqrt[6]{108}\sqrt[3]{2} = \sqrt[6]{4 \cdot 108} = \sqrt[6]{432}$ etc. However, there are only a finite number of different irrationalities generated in this fashion since every such irrationality can be expressed as $k\sqrt[6]{2^m 3^n}$ where $2^m 3^n < 2^6 3^6$. Thus there exists a vector space $V = \{a + b\sqrt[3]{2} + c\sqrt{3} + d\sqrt[6]{108} + e\sqrt[6]{432} + \ldots | a, b, c, d, e, \ldots \in R$ over R of dimension n over R such that $1, \alpha, \ldots, \alpha^n$ are linearly dependent in V. Thus there exist $c_1, \ldots, c_{n+1}$, not all zero, in R such that $c_1 + c_2\alpha + \ldots + c_{n+1}\alpha^n = 0$.

† It is actually four since 1, $\sqrt{2}$, $\sqrt{3}$, and $\sqrt{6}$ are linearly independent over R. But we do not need to know this to achieve our desired result.

In general, we can conclude that any linear combination, with coefficients in R, of radicals of the form $\sqrt[m]{a}$ satisfies a polynomial equation with rational coefficients.

PROBLEMS FOR FURTHER INVESTIGATION

1. The *dot*, *scalar*, or *inner* product of two vectors $\xi = (x_1, \ldots, x_n)$ and $\eta = (y_1, \ldots, y_n)$ of $V_n(F)$ is defined to be $\xi \cdot \eta = \sum_{i=1}^{n} x_i y_i$. Investigate the properties of this inner product. [W1]
2. The inner product, $\xi \cdot \eta$, described in problem 1 is not in $V_n(F)$ but in F. Consider the problem of defining a multiplication of vectors of $V_n(F)$ to produce another vector, beginning with a consideration of the complex numbers and the quaternions considered as vectors. [A1], [B4], [D3].
3. Since V is an Abelian group under addition we can, for any subspace R of V, consider V/R. Consider these quotient spaces in detail when $V = V_n(F)$ and describe them geometrically when $n = 2$ or 3. [H1]

REFERENCES FOR FURTHER READING

[B1], [B4], [H1], [K2], [M6].

12

Polynomials

1. Definitions

In previous chapters we have used the concept of a polynomial informally in examples just as we used the concept of a real number before the formal treatment of real numbers in Chapter 8. In this chapter we will give a formal definition of a polynomial and discuss properties of polynomials and polynomial rings.

One way of considering polynomials is to take an integral domain and to "adjoin" to it a symbol (indeterminate) "x". We define $x^1 = x, x^n = x \cdot x^{n-1}$ as additional symbols and write, for example, $x^2 + 3x + 1$ without any concern for the meaning of x^2, $3x$, $3x + 1$ etc. Difficulties arise, however, when we, for example "take $x = 2$" in $x^2 + 3x + 1$.

To avoid the difficulty of regarding x as an indeterminate and yet being able to replace x by a number, we may consider vectors $(a_0, a_1, a_2, \ldots, a_n, 0, 0, \ldots)$ with an infinite number of entries but with only a finite number of nonzero entries. Thus, for example, $-1 + 2x - 3x^2 + x^5$ is replaced by $(-1, 2, -3, 0, 0, 1, 0, 0, \ldots)$.

Definition 1. *Let D be an integral domain. Then a* ***polynomial*** *over D is an infinite sequence $(a_0, a_1, \ldots, a_n, 0, 0, \ldots)$ of elements of D all but a finite number of which are zero.*

By the ***sum*** $(a_0, a_1, \ldots, a_n, 0, 0, \ldots) + (b_0, b_1, \ldots, b_m, 0, 0, \ldots) = (b_0, b_1, \ldots, b_m, 0, 0, \ldots) + (a_0, a_1, \ldots, a_n, 0, 0, \ldots)$ *of two polynomials* $(a_0, a_1, \ldots, a_n, 0, 0, \ldots)$ *and* $(b_0, b_1, \ldots, b_m, 0, 0, \ldots)$ $(m \geq n)$ *we mean the polynomial*

$$(a_0 + b_0, a_1 + b_1, \ldots, a_n + b_n, b_{n+1}, \ldots, b_m, 0, 0, \ldots)$$

By the ***scalar product***, $a(a_0, a_1, \ldots, a_n, 0, 0, \ldots)$, *of the polynomial* $(a_0, a_1, \ldots, a_n, 0, 0, \ldots)$ *by an element* $a \in D$ *we mean the polynomial*

$$(aa_0, aa_1, \ldots, aa_n, 0, 0, \ldots)$$

We also define $(a_0, a_1, \ldots, a_n, 0, 0, \ldots)a = a(a_0, a_1, \ldots, a_n, 0, 0, \ldots)$.

By the **product**, $(a_0, a_1, \ldots, a_n, 0, 0, \ldots)(b_0, b_1, \ldots b_m, 0, 0, \ldots)$, *of two polynomials* $(a_0, a_1, \ldots, a_n, 0, 0, \ldots)$ *and* $(b_0, b_1, \ldots, b_m, 0, 0, \ldots)$ *we mean the polynomial*

$$(a_0b_0, a_0b_1 + a_1b_0, a_0b_2 + a_1b_1 + a_2b_0, \ldots, a_nb_m, 0, 0, \ldots)$$

where the ith entry is

$$\sum_{j+k=i-1} a_jb_k$$

Finally, if $\alpha = (a_0, a_1, \ldots, a_n, 0, 0, \ldots)$ *and* $c \in D$, *we define the* **value**, α_c, *of* α *at* c *to be* $\alpha_c = a_0 + a_1c + \ldots + a_nc^n \in D$.

If now we write $x = (0, 1, 0, 0, \ldots)$ we have $x^2 = (0, 0, 1, 0, 0, \ldots)$, $x^3 = (0, 0, 0, 1, 0, 0, \ldots)\ldots$ so that $(a_0, a_1, a_2, \ldots, a_n, 0, 0, \ldots) = a_0e + a_1x + a_2x^2 + \ldots + a_nx^n$ where $e = (1, 0, 0, \ldots)$. Thus all we have done is to prescribe, in vector notation, the usual rules for operating with polynomials and for finding $f(c)$ when a polynomial $f(x)$ is given. Thus our customary notation for polynomials that we will also use is simply a more convenient (or at least more familiar) way of writing down the sequence $a_0, a_1, \ldots, a_n$; certainly we do not need to worry about the meaning of "x". What we do need to show is embodied in the following theorem.

Theorem 1. If $\alpha = (a_0, a_1, \ldots, a_n, 0, 0, \ldots)$, $\beta = (b_0, b_1, \ldots, b_m, 0, 0, \ldots)$, $\gamma = \alpha + \beta$, $\delta = a\alpha\ (a \in D)$, and $\rho = \alpha\beta$, then $\gamma_c = \alpha_c + \beta_c$, $\delta_c = a\alpha_c$, and $\rho_c = \alpha_c\beta_c$. Briefly, if P is the set of all polynomials over D, then $\alpha \in P \to \alpha_c \in D$ is a homomorphism of P into D (where the correspondence is preserved under addition, scalar multiplication, and multiplication).

Note that this result assures us that computations performed on polynomials over D can be translated into computations over D. This, of course, is what makes polynomials so useful since, for example, $x^2 - 4 = (x + 2)(x - 2)$ means that $5^2 - 4 = (5 + 2)(5 - 2)$, $6^2 - 4 = (6 + 2)(6 - 2)$ etc.

Proof. We prove only the first part of the theorem leaving the other two parts as exercises for the student. We have, for $m \geq n$,

$$\begin{aligned}
\alpha_c + \beta_c &= (a_0 + a_1c + \ldots + a_nc^n) + (b_0 + b_1c + \ldots + b_mc^m) \\
&= (a_0 + b_0) + (a_1 + b_1)c + \ldots + (a_n + b_n)c^n + b_{n+1}c^{n+1} \\
&\qquad + \ldots + b_mc^m \\
&= (\alpha + \beta)_c = \gamma_c
\end{aligned}$$

From now on we will use the customary notation for polynomials: $a_0 + a_1x + \ldots + a_nx^n$.

Definition 2. *A polynomial* $f(x) = a_0 + a_1x + \ldots + a_nx^n$ $(a_n \neq 0)$ *over* D *is said to have* ***degree*** n. *We write* ***deg*** $f(x) = n$. *If* $f(x) = 0 \in D$ *we make the convention that deg* $f(x) = 0$.

Theorem 2. If $f(x) \neq 0$ and $g(x) \neq 0$ are two polynomials over an integral domain, then *deg* $[f(x)g(x)] =$ *deg* $f(x) +$ *deg* $g(x)$.

Proof. If $f(x) = a_0 + a_1x + \ldots + a_nx^n$ and $g(x) = b_0 + b_1x + \ldots + b_mx^m$ $(a_n \neq 0, b_m \neq 0)$ we have

$$f(x)g(x) = a_0b_0 + (a_0b_1 + a_1b_0)x + \ldots + a_nb_mx^{n+m}$$

Since $a_n \neq 0$ and $b_m \neq 0$, it follows that $a_nb_m \neq 0$ since a_n and b_m are elements of an integral domain. Hence *deg* $[f(x)g(x)] = n + m$.

Exercises 12.1

1. Add the following polynomials:
(a) $(2, -1, 0, 3, 0, 0, \ldots)$ and $(0, 2, 1, -3, 5, 0, 0, \ldots)$
(b) $(-1, \sqrt{2}, 1, \pi, 0, \ldots)$ and $(-1, \sqrt{3}, -1, 2, 0, 0, \ldots)$
(c) $(-3, 0, 0, 1, 0, 0, \ldots)$ and $(3, 0, 0, -1, 0, 0, \ldots)$
(d) $(\sqrt{2}, 1, -1, 0, 2, 0, 0, \ldots)$ and $(-\sqrt{2}, -1, 1, 0, -2, 0, 0, \ldots)$

2. Multiply the following polynomials:
(a) $(-1, 0, 2, 0, 0, \ldots)$ and $(0, 2, -1, 0, 0, \ldots)$
(b) $(2, 1, 0, 0, \ldots)$ and $(3, 1, 0, 0, \ldots)$
(c) $(1, 1, 1, 0, 0, \ldots)$ and $(1, -1, 0, 0, \ldots)$
(d) $(-1, 1, -1, 1, 0, 0, \ldots)$ and $(1, 1, 0, 0, \ldots)$

3. Illustrate Theorem 1 by taking $c = 2$ in problems 1a, 1c, 2a, and 2c.

4. Complete the proof of Theorem 1.

2. The Division Algorithm

It is easy to show that, for any integral domain D, $D[x] = \{a_0 + a_1x + \ldots + a_nx^n | n \in N, a_i \in D, i = 1, \ldots, n\}$ is also an integral domain whose units are the units of D. Thus two polynomials $f(x)$ and $g(x)$ of $D[x]$ are associates if and only if $f(x) = ug(x)$ where u is a unit of D. (See Section 5.3.)

In the remaining part of this chapter we will consider only the case when D is a field although some of the results we will state will be true when D is simply an integral domain. (Recall, in this connection, that every nonzero element of a field is a unit of the field so that any nonzero multiple of a polynomial over a field is an associate of the polynomial.)

Theorem 3. (*the division algorithm*) Let F be a field. If $f(x), g(x) \in F[x]$ and $g(x) \neq 0$, then there exist unique $q(x), r(x) \in F[x]$ such that

$$f(x) = q(x)g(x) + r(x)$$

where $deg\ r(x) < deg\ g(x)$ (or, if $g(x) \in F, r(x) = 0$).

Proof. Let $f(x) = a_0 + a_1x + \ldots + a_nx^n$ and $g(x) = b_0 + b_1x + \ldots + b_mx^m$ ($b_m \neq 0$). If $deg\ f(x) < deg\ g(x)$ we have

$$f(x) = 0 \cdot g(x) + f(x)$$

so that we have the desired result in this case.

Furthermore, if $f(x) \neq 0$ but $deg\ f(x) = 0$ so that $f(x) = a_0$ we have, for $deg\ g(x) = m = 0$,

$$f(x) = a_0 = \frac{a_0}{b_0} b_0 + 0$$

and, for $m > 0$,

$$f(x) = 0 \cdot g(x) + a_0$$

so that our result holds if $deg\ f(x) = n = 0$. We proceed with an inductive proof by assuming that the theorem is true when $deg\ f(x) < n$ and showing that then it is also true when $deg\ f(x) = n$. (See Theorem 5.11.† To this end take $n \geq m$ and $n > 0$ and let

$$(1) \qquad f_1(x) = f(x) - \frac{a_n}{b_m} x^{n-m}g(x)$$

Then $deg\ f_1(x) < n$ so that by our induction hypothesis we may write

$$(2) \qquad f_1(x) = q_1(x)g(x) + r(x)$$

where $deg\ r(x) < deg\ g(x)$ (or $r(x) = 0$ if $g(x) \in F$). But now we may write (1) with the aid of (2) as

$$f(x) - \frac{a_n}{b_m} x^{n-m}g(x) = q_1(x)g(x) + r(x)$$

so that

$$f(x) = \left[\frac{a_n}{b_m} x^{n-m} + q_1(x)\right]g(x) + r(x) = q(x)g(x) + r(x)$$

as desired.

† Slightly modified. We leave it to the student to show that if "$1 \in S$" in Theorem 5.11 is replaced by "$0 \in S$", the conclusion becomes $S = N \cup \{0\}$.

Now suppose that, also, $f(x) = q'(x)g(x) + r'(x)$ where $deg\, r'(x) < deg\, g(x)$. Then

$$q'(x)g(x) + r'(x) = q(x)g(x) + r(x)$$

so that

$$g(x)[q'(x) - q(x)] = r(x) - r'(x)$$

But $deg\, [r(x) - r'(x)] < deg\, g(x)$ so that, by Theorem 2, $q'(x) - q(x) = 0$. Hence $q'(x) = q(x)$ and $r'(x) = r(x)$. (If $g(x) \in F$, then $r(x) = r'(x) = 0$.)

Definition 3. *The polynomial $q(x)$ in the division algorithm is called the* ***quotient*** *of $f(x)$ upon division by $g(x)$ and the polynomial $r(x)$ is called the* ***remainder.***

Corollary 1. (*the remainder theorem*) The remainder when a polynomial $f(x)$ is divided by $x - a$ is $f(a)$.

Proof. If $g(x) = x - a$ in the division algorithm we have

$$f(x) = q(x)(x - a) + r(x)$$

where $\deg r(x) < 1$. Hence $r(x) = b \in F$. But then $f(a) = q(x)(a - a) + b = b$ as desired.

Corollary 2. (*the factor theorem*) A polynomial $f(x)$ is divisible by $x - a$ if and only if $f(a) = 0$.

Proof. If $f(a) = 0$, then $x - a$ divides $f(x)$ by Corollary 1. On the other hand, if $x - a$ divides $f(x)$, we have $f(x) = q(x)(x - a)$ so that $f(a) = q(a)(a - a) = 0$.

Another useful by-product of the division algorithm is the process known as *synthetic division.* This process provides an easy way of finding the quotient $q(x)$ and the remainder, r, when a polynomial $f(x) \in F[x]$ is divided by the polynomial $x - c \in F[x]$. Let $f(x) = a_0 + a_1x + \dots + a_{n-1}x^{n-1} + a_nx^n$ and $q(x) = b_0 + b_1x + \dots + b_{n-1}x^{n-1}$. Then

$$\begin{aligned} f(x) &= (x - c)q(x) + r \\ &= (r - cb_0) + (b_0 - cb_1)x + (b_1 - cb_2)x^2 + \cdots \\ &\qquad + (b_{n-2} - cb_{n-1})x^{n-1} + b_{n-1}x^n \end{aligned}$$

When we equate coefficients in the two representations of $f(x)$ we have

$$a_n = b_{n-1},\ a_{n-1} = b_{n-2} - cb_{n-1},\ \dots,\ a_1 = b_0 - cb_1,\ a_0 = r - cb_0$$

For computational purposes this work may be arranged as follows:

a_n	a_{n-1}	$\cdots$	a_1	a_0	$\lfloor c$
	cb_{n-1}	$\cdots$	cb_1	cb_0	
$b_{n-1} = a_n$	$b_{n-2} = a_{n-1} + cb_{n-1}$		$b_0 = a_1 + cb_1$	$r = a_0 + cb_0$	

Example. Find the quotient and remainder when $4x^4 - 3x^2 - x + 2$ is divided by $x + 2$ $(= x - (-2))$.

4	0	−3	−1	2	$\lfloor -2$
	−8	16	−26	54	
4	−8	13	−27	56	

The quotient is $4x^3 - 8x^2 + 13x - 27$ and the remainder is 56.

Exercises 12.2

1. Find the quotient and remainder when
   (a) $5x^3 - 3x^2 + 10x - 1$ is divided by $x - 2$.
   (b) $x^5 - 4x^2 - x + 2$ is divided by $x + 1$.
   (c) $x^3 + ix + 2$ is divided by $x - i$.
   (d) $2x^3 - 3x + 1$ is divided by $x + \sqrt{2}$.
2. If $f(x) = 3x^3 + 2x^2 - 1$, find $f(3)$, $f(-1)$, and $f(i)$.
3. In $J_5[x]$, exhibit the quotient and remainder when the polynomial $3x^3 + 2x^2 - 4x - 2$ is divided by $x^2 - x + 1$.

3. Greatest Common Divisor

Definition 4. *If $f(x)$, $g(x)$, and $q(x) \in F[x]$ and $f(x) = q(x)g(x)$, we say that $g(x)$* ***divides*** *$f(x)$ and that $g(x)$ is a* ***divisor*** *of $f(x)$. We write $g(x)|f(x)$.*

Thus, for example, if $F = R$ we have $(x - 1) \mid (2x - 2)$ since $2x - 2 = 2(x - 1)$ and, also, $(2x - 2) \mid (x - 1)$ since $x - 1 = \frac{1}{2}(2x - 2)$. In general, if $g(x) \mid f(x)$, $ag(x) \mid f(x)$ for any $a \neq 0 \in F$, i.e., if $g(x) \mid f(x)$, any associate of $g(x)$ also divides $f(x)$. In particular, if $g(x) = a_0 + a_1x + \ldots + a_nx^n$ $(a_n \neq 0)$ and $g(x) \mid f(x)$, then $a_n^{-1}g(x) \mid f(x)$. We note that $a_n^{-1}g(x) = a_n^{-1}a_0 + a_n^{-1}a_1x + \ldots + ex^n$ where the coefficient of x^n is the unity element of F and write $ex^n = x^n$.

Definition 5. *A polynomial $g(x) \in F[x]$ is called a* **monic polynomial** *if the coefficient of the highest power of x in $g(x)$ is the unity element of F.*

Definition 6. *We say that $d(x)$ is the* **greatest common divisor (g.c.d.)** *of two polynomials $f(x)$ and $g(x)$ if*
(1) $d(x) \mid f(x)$ and $d(x) \mid g(x)$;
(2) $h(x) \mid f(x)$ and $h(x) \mid g(x)$ implies $h(x) \mid d(x)$;
(3) *$d(x)$ is a monic polynomial.*

We write $d(x) = (f(x), g(x))$. If $d(x) = e$, the unity element of F, we say that $f(x)$ and $g(x)$ are ***relatively prime.***

Theorem 4. (*the Euclidean algorithm*) Any two nonzero polynomials $f(x)$ and $g(x) \in F[x]$ have a unique g.c.d., $d(x) \in F[x]$.

Proof. The proof follows the same pattern as that of the corresponding theorem for the integers (Section 5.6). We give only an outline of the proof here, leaving the details to be filled in by the student.

By the division algorithm we obtain the sequence of equalities

$$
\begin{array}{rlr}
f(x) &= q(x)g(x) + r(x) & deg\ r(x) < deg\ g(x) \\
g(x) &= q_1(x)r(x) + r_1(x) & deg\ r_1(x) < deg\ r(x) \\
r(x) &= q_2(x)r_1(x) + r_2(x) & deg\ r_2(x) < deg\ r_1(x) \\
r_1(x) &= q_3(x)r_2(x) + r_3(x) & deg\ r_3(x) < deg\ r_2(x) \\
\vdots & \quad \vdots \qquad\quad \vdots & \vdots \qquad\quad \vdots \\
r_{n-2}(x) &= q_n(x)r_{n-1}(x) + r_n(x) & deg\ r_n(x) < deg\ r_{n-1}(x) \\
r_{n-1}(x) &= q_{n+1}(x)r_n(x) &
\end{array}
$$

Now we observe that $(f(x), g(x)) = (g(x), r(x)) = (r(x), r_1(x)) = \ldots = (r_{n-2}(x), r_{n-1}(x)) = (r_{n-1}(x), r_n(x)) = r_n(x)$.

To prove uniqueness we observe that if $d(x)$ and $d'(x)$ are both g.c.d.'s of $f(x)$ and $g(x)$ we have $d(x) \mid d'(x)$ and $d'(x) \mid (dx)$. Thus $d(x) = h'(x)d'(x)$ and $d'(x) = h(x)d(x)$. Hence $d(x) = h'(x)h(x)d(x)$ so that $h'(x)h(x) = e$, the unity element of F. But then $h'(x) = b' \in F$ and $h(x) = b \in F$. Therefore $d'(x) = bd(x)$ and since $d'(x)$ and $d(x)$ are both monic polynomials it follows that $b = e$ and hence that $d'(x) = d(x)$.

Although our definition of the g.c.d. of two polynomials yields a unique g.c.d. (because of the monic requirement), it is frequently convenient to call any associate of the unique g.c.d. also a g.c.d. We shall sometimes do so when no confusion can arise and will find this practice useful in computing the g.c.d. as illustrated in the following example.

Example 1. Find the g.c.d. of $f(x) = x^5 - x^4 - 6x^3 - 2x^2 - x - 3$ and $g(x) = x^3 - 3x - 2 \in R[x]$.

We first write $x^5 - x^4 - 6x^3 - 2x^2 - x - 3 = (x^2 - x - 3)(x^3 - 3x - 2) + (-3x^2 - 12x - 9)$ but we now consider $-\frac{1}{3}(-3x^2 - 12x - 9) = x^2 + 4x + 3$ instead of $-3x^2 - 12x - 9$ in order to avoid fractions. Thus we write

$$x^3 - 3x - 2 = (x - 4)(x^2 + 4x + 3) + (10x + 10)$$

Again we consider $(\frac{1}{10})(10x + 10) = x + 1$ instead of $10x + 10$ and write

$$x^2 + 4x + 3 = (x + 3)(x + 1) + 0$$

Thus $x + 1$ is *a* g.c.d. and, since it is monic, it is *the* g.c.d.

Theorem 5. Let $d(x)$ be the g.c.d. of two polynomials $f(x)$ and $g(x) \in F[x]$. Then there exist polynomials $m(x)$ and $n(x) \in F[x]$ such that

$$d(x) = m(x)g(x) + n(x)f(x)$$

We leave it to the reader to supply a proof along the lines of the proof of Theorem 5.15.

Example 2. In showing that $(x^5 - x^4 - 6x^3 - 2x^2 - x - 3, x^3 - 3x - 2) = x + 1$, we obtained the following sequence of equalities:

$$\begin{aligned} x^5 - x^4 - 6x^3 - 2x^2 - x - 3 &= (x^2 - x - 3)(x^3 - 3x - 2) \\ &\quad + (-3x^2 - 12x - 9) \\ x^3 - 3x - 2 &= (x - 4)(x^2 + 4x + 3) + (10x + 10) \\ x^2 + 4x + 3 &= (x + 3)(x + 1) + 0 \end{aligned}$$

Hence

$$\begin{aligned} 10x + 10 &= \underline{(x^3 - 3x - 2)} - (x - 4)\underline{(x^2 + 4x + 3)} \\ -30x - 30 &= -3\underline{(x^3 - 3x - 2)} - (x - 4)\underline{(-3x^2 - 12x - 9)} \\ -30x - 30 &= -3\underline{(x^3 - 3x - 2)} - (x - 4)[\underline{(x^5 - x^4 - 6x^3 - 2x^2 - x - 3)} \\ &\quad - (x^2 - x - 3)\underline{(x^3 - 3x - 2)}] \\ &= [(x - 4)(x^2 - x - 3) - 3]\underline{(x^3 - 3x - 2)} \\ &\quad + (4 - x)\underline{(x^5 - x^4 - 6x^3 - 2x^2 - x - 3)} \\ &= (x^3 - 5x^2 + x + 9)\underline{(x^3 - 3x - 2)} \\ &\quad + (4 - x)\underline{(x^5 - x^4 - 6x^3 - 2x^2 - x - 3)} \end{aligned}$$

and

$$x + 1 = (-\tfrac{1}{30}x^3 + \tfrac{1}{6}x^2 - \tfrac{1}{30}x + \tfrac{3}{10})\underline{(x^3 - 3x - 2)}$$
$$+ (\tfrac{1}{30}\,x - \tfrac{2}{15})\,\underline{(x^5 - x^4 - 6x^3 - 2x^2 - x - 3)}$$

(Notice how we again avoided the use of fractions until the last step.)

Corresponding to the notion of a prime integer we have the notion of a prime or irreducible polynomial.

Definition 7. *$f(x) \in F[x]$ is said to be a **prime** or **irreducible** polynomial over F if and only if the only divisors of $f(x)$ are its associates and the units of F.*

Example 3. All of the elements of F (except 0) are irreducible polynomials (of degree 0); $2x - 2$ is irreducible over R^*; it has the divisors $ax - a$ $(a \neq 0 \in R^*)$ and $a \neq 0 \in R^*$ but these are either associates of $2x - 2$ or units of R^*; $x^2 - 2$ is irreducible over R but reducible $[=(x + \sqrt{2})(x - \sqrt{2})]$ in R^*; $x^2 + 1$ is irreducible over R^* but reducible over C $[=(x + i)(x - i)]$ and over J_2 $[=(x + 1)(x + 1)]$.

Once again we ask the student to prove some theorems on divisibility of polynomials utilizing, if necessary, the procedures used in the corresponding theorems for integers (Section 5.7).

Theorem 6. If $p(x)$ is an irreducible polynomial over F and if $p(x) \mid f(x)g(x)$ for $f(x), g(x) \in F[x]$, then $p(x) \mid f(x)$ or $p(x) \mid g(x)$.

Corollary. If $p(x)$ is an irreducible polynomial over F and if $p(x) \mid f_1(x) f_2(x) \dots f_n(x)$ for $f_1(x), f_2(x), \dots, f_n(x) \in F[x]$, then $p(x) \mid f_i(x)$ for some $i = 1, 2, \dots, n$.

Theorem 7. If $(f(x), g(x)) = e$, the unity element of F, and if $f(x) \mid g(x) h(x)$ for $f(x), g(x), h(x) \in F[x]$, then $f(x) \mid h(x)$.

We now are in a position to prove the unique factorization theorem for polynomials.

Theorem 8. Suppose $f(x) \in F[x]$ is of degree > 0. Then $f(x)$ can be expressed uniquely, except for order of factors, as an element of F times a product of monic irreducible polynomials in $F[x]$.

Proof. If $f(x)$ is irreducible, the result is trivial. In particular, if $deg\, f(x) = 1$, $f(x)$ is irreducible. Suppose now that $f(x)$ is reducible and $deg\, f(x) = n > 1$. Then $f(x) = g(x)h(x)$ where $g(x)$ and $h(x)$ are polynomials of degree less than $deg\, f(x)$. We make the induction hypothesis that the decomposition is possible for polynomials of degree less than n. Then

$$g(x) = cp_1(x)p_2(x) \dots p_r(x) \quad \text{and} \quad h(x) = c'p'_1(x)p'_2(x) \dots p'_s(x)$$

where c and c' are in F and the $p_i(x)$ and $p'_i(x)$ are monic irreducible polynomials of $F[x]$. Then

$$f(x) = g(x)h(x) = cc'p_1(x) \ldots p_r(x)p'_1(x) \ldots p'_s(x)$$

and our induction is completed as to the existence of such a decomposition.

Now suppose that we have two such decompositions. Obviously this is impossible if $deg\ f(x) = 1$ so we assume that $deg\ f(x) = n > 1$ and that

$$f(x) = cp_1(x)p_2(x) \ldots p_n(x) = dq_1(x)q_2(x) \ldots q_m(x)$$

where $c, d \in F$ and the $p_i(x)$ and $q_i(x)$ are monic irreducible polynomials of $F[x]$. Clearly, $c = d$ and also, by the corollary to Theorem 6, $p_1(x)|q_i(x)$ for some $i = 1, 2, \ldots, m$. Therefore, since $p_1(x)$ and $q_i(x)$ are monic and irreducible, $p_1(x) = q_i(x)$. Dividing by $p_1(x)$ and c we obtain

$$f_1(x) = p_2(x) \ldots p_n(x) = q_1(x) \ldots q_{i-1}(x)q_{i+1}(x) \ldots q_m(x)$$

We now make the induction hypothesis that the decomposition is unique for polynomials of degree less than n. Since $f_1(x)$ is such a polynomial, its decomposition is unique and hence the decomposition of $f(x)$ is unique.

Exercises 12.3

1. Find $(f(x), g(x))$ over the indicated field, F, of coefficients and express it in the form $m(x)g(x) + n(x)f(x)$ for $m(x), n(x) \in F[x]$:
   (a) $f(x) = x^5 - x^4 - 6x^3 - 2x^2 + 5x + 3$, $g(x) = x^3 - 3x - 2$; $F = R$
   (b) $f(x) = x^4 - 5x^3 + 6x^2 + 4x - 8$, $g(x) = x^3 - x^2 - 4x + 4$; $F = R$
   (c) $f(x) = x^4 - 4ix + 3$, $g(x) = x^3 - i$; $F = C$
2. In $R[x]$ find $(f(x), g(x))$ when
   (a) $f(x) = x^4 - x^3 - 3x^2 + x + 2$, $g(x) = x^3 - x^2 - 5x + 2$
   (b) $f(x) = x^4 + x^3 + 2x^2 + x + 1$, $g(x) = x^3 - 1$
3. In $J_3[x]$ find $(f(x), g(x))$ when
   (a) $f(x) = x^5 + 2x^3 + x^2 + 2x$, $g(x) = x^4 + x^3 + x^2$
   (b) $f(x) = x^3 + 2x^2 + 2x + 1$, $g(x) = x^2 + 2$
4. In $R[x]$ determine c so that $(f(x), g(x))$ is linear and then find $(f(x), g(x))$ if
   (a) $f(x) = x^3 + cx^2 - x + 2c$, $g(x) = x^2 + cx - 2$
   (b) $f(x) = x^2 + (c - 6)x + 2c - 1$, $g(x) = x^2 + (c + 2)x + 2c$
5. Prove Theorem 5.
6. Prove Theorem 6.
7. Prove the corollary to Theorem 6.
8. Prove Theorem 7.

4. Zeros of a Polynomial

Definition 8. *If* $f(x) \in F[x]$, $a \in F$, and $f(a) = 0$ *we say that* a *is a zero of* $f(x)$.

The determination of the zeros of a polynomial is, in general, a complicated task. Certain results are known for polynomials of any degree over an arbitrary field; other results are known for polynomials of certain degrees (in particular, degrees 1, 2, 3, and 4) over arbitrary fields; and still other results are dependent upon the particular field under consideration (especially for R, R^*, and C).

The student is, of course, familiar with the results concerning polynomials of degree 1 and 2 (linear and quadratic). Here we will first consider two results about polynomials that are independent of the field and then discuss some results for the cases when $F = C$, R^*, R, and J_p respectively.

Theorem 9. If $f(x) \in F[x]$ and $deg\, f(x) = n > 1$, then $f(x)$ has at most n zeros in F.

Proof. If $n = 1$, $f(x) = a_0 + a_1x$ $(a_1 \neq 0)$ and $f(x)$ has the one zero, $-a_0/a_1$. Thus the theorem is true for $n = 1$. Now suppose $deg\, f(x) > 1$. If $f(x)$ has a zero, r_1, the factor theorem gives us $f(x) = (x - r_1)q(x)$ where $q(x)$ is of degree $n - 1$. Clearly, every zero of $q(x)$ is a zero of $f(x)$. On the other hand, $f(x)$ has no other zeros than r_1 and the zeros of $q(x)$. For if $f(a) = 0$, then $(a - r_1)q_1(a) = 0$. Hence $a = r_1$ or $q_1(a) = 0$ since a field has no divisors of zero.

If we now make the induction hypothesis that the theorem is true for polynomials of degree less than n, we see that $q(x)$ has at most $n - 1$ zeros in F so that, by the above argument, $f(x)$ has at most $1 + (n - 1) = n$ zeros in F.

The proof of the following theorem closely parallels the proof of Theorem 8 and is left as an exercise for the student.

Theorem 10. If $f(x) = a_0 + a_1x + \ldots + a_nx^n \in F[x]$ has the n distinct zeros $r_1, r_2, \ldots, r_n \in F$, then $f(x)$ can be written uniquely, except for order, as the product of a_n and n linear monic polynomials. Specifically, $f(x) = a_n(x - r_1)(x - r_2) \ldots (x - r_n)$.

Note. If $f(x) = (x - r)^m q(x)$ where $q(r) \neq 0$ we say that $f(x)$ has a zero, r, of *multiplicity* m. If we list each zero of a polynomial as many times as its multiplicity we may modify Theorem 10 to eliminate the word "distinct." Thus if $f(x)$ has the zero 2 of multiplicity 3, the zero -1 of multiplicity 2, the zero 4 of multiplicity 1, and no other zeros, we know that $f(x) = a(x - 2)^3 (x + 1)^2(x - 4)$.

In the case when $F = C$ we recall that we have already remarked (Theorem 10.3) that the complex numbers form an algebraically closed field. That is, if $f(x) \in C[x]$, $f(x)$ has a zero in C.

Theorem 11. If $f(x) = a_0 + a_1x + \ldots + a_nx^n \in C[x]$, then $f(x) = a_n (x - r_1)(x-r_2) \ldots (x - r_n)$ where $r_i \in C$ for $i = 1, 2, \ldots, n$.

Proof. If $n = 1$, the theorem is certainly true. If $n > 1$, we know by Theorem 10.3 that $f(x)$ has a zero $r_1 \in C$. By the factor theorem, $f(x) = (x - r_1)q(x)$. Now we proceed as in the proof of Theorem 9.

We now rephrase Theorem 7.15.

Theorem 12. If $f(x) \in R^*[x]$ has the zero $a + bi$ $(a, b \in R^*, b \neq 0)$, then it also has the *conjugate zero* $a - bi$.

Theorem 13. If $f(x) \in R^*[x]$, then $f(x)$ can be written as a product of a real number times a product of monic irreducible quadratic factors and monic linear factors in $R^*[x]$.

Proof. We may consider $f(x)$ as a polynomial of $C[x]$ since $C \supset R^*$. In $C[x]$, then, $f(x)$ factors into linear factors (Theorem 11). Now we use Theorem 12 to observe that every linear factor $x - (a + bi)$ corresponding to a zero of the form $a + bi$ $(a, b \in R^*, b \neq 0)$ can be paired with a linear factor $x - (a - bi)$. Our proof is then completed by observing that $[x - (a + bi)][x - (a - bi)] = (x - a)^2 + b^2$ is quadratic and clearly irreducible over R^*.

Our next two results concern factorization of polynomials in $R[x]$.

Theorem 14. Let c/d where $(c, d) = 1$ be a rational zero of $f(x) = a_0 + a_1x + \dots + a_nx^n \in I[x]$. Then $c|a_0$ and $d|a_n$.

Proof. Since $a_0 + a_1(c/d) + \dots + a_{n-1}(c/d)^{n-1} + a_n(c/d)^n = 0$ we have

$$a_0d^n + a_1cd^{n-1} + \dots + a_{n-1}c^{n-1}d + a_nc^n = 0$$

Hence $d(a_0d^{n-1} + a_1cd^{n-2} + \dots + a_{n-1}c^{n-1}) + a_nc^n = 0$ and $d|a_nc^n$. But $(c^n, d) = 1$ (problem 5, Exercises 5.7) and hence $d|a_n$. Similarly, $a_0d^n + c(a_1d^{n-1} + \dots + a_nc^{n-1}) = 0$ so that $c|a_0d^n$. Again $(c, d^n) = 1$ and hence $c|a_0$.

The proof of the following corollary is left as an exercise for the student.

Corollary. If $f(x) = a_0 + a_1x + \dots + a_{n-1}x^{n-1} + x^n \in I[x]$ has a root $r \in R$, then $r \in I$ and $r|a_0$.

Note that this corollary gives us a simple way of proving the irrationality of certain real numbers. Thus, for example, $\sqrt{3}$ and $\sqrt[3]{5}$ are irrational because they are zeros of the polynomials $x^2 - 3$ and $x^3 - 5$ respectively and, by the use of the corollary, it is easily checked that these polynomials have no rational roots.

Example. Find the decomposition of the polynomial $f(x) = 9x^4 + 12x^3 + 10x^2 + x - 2$ over R and over C.

The possible rational roots are, by Theorem 14, ± 1, ± 2, $\pm 1/3$, $\pm 1/9$, $\pm 2/3$, and $\pm 2/9$. Using synthetic division we find that none of ± 1, ± 2 are zeros but that $1/3$ is a zero:

$$\begin{array}{rrrrr|l} 9 & 12 & 10 & 1 & -2 & \underline{1/3} \\ & 3 & 5 & 5 & 2 & \\ \hline 9 & 15 & 15 & 6 & 0 & \end{array}$$

Hence $f(x) = (x - 1/3)(9x^3 + 15x^2 + 15x + 6) = 3(x - 1/3)(3x^3 + 5x^2 + 5x + 2)$. Now the possibilities for the zeros of $g(x) = 3x^3 + 5x^2 + 5x + 2$ are $\pm 1/3$, $\pm 1/9$, $\pm 2/3$, and $\pm 2/9$. Using synthetic division again we find that $\pm 1/3$, $\pm 1/9$, and $2/3$ are not zeros of $g(x)$ but that $-2/3$ is a zero:

$$\begin{array}{rrrr|l} 3 & 5 & 5 & 2 & \underline{-2/3} \\ & -2 & -2 & -2 & \\ \hline 3 & 3 & 3 & 0 & \end{array}$$

Thus $f(x) = 3(x - 1/3)(x + 2/3)(3x^2 + 3x + 3) = 9(x - 1/3)(x + 2/3)(x^2 + x + 1)$. It is easy to see that $x^2 + x + 1$ is irreducible in R (and, in fact, in R^*) so that we have achieved a decomposition into irreducible factors in R. The decomposition in C, then, is

$$f(x) = 9(x - 1/3)(x - 2/3)\left(x - \frac{-1 + \sqrt{3}i}{2}\right)\left(x + \frac{1 + \sqrt{3}i}{2}\right)$$

We turn now to a brief consideration of the problem of finding zeros of polynomials over finite fields, J_p. Here, in contrast to the situation for infinite fields, we may actually examine *all* of the possibilities for zeros. For example, in J_2 the only possible zeros for any polynomial are 0 and 1. In particular then, $x^2 + x + 1$ is irreducible in J_2 since $0^2 + 0 + 1 \neq 0$ and $1^2 + 1 + 1 = 1 \neq 0$.

We may exploit this situation in seeking zeros or factors of polynomials in $I[x]$ or $R[x]$ since the equality of two integers in I, $a = b$, certainly implies their equality in J_p, $a \equiv b \pmod{p}$ (although, of course, not conversely). Thus, for example, $2 + 3 = 5, 2 \times 3 = 6$ and $x^2 + 5x + 6 = (x + 2)(x + 3)$ are identities modulo m for any positive integer m.

For example, suppose we wish to know whether or not $27x^2 + 159x - 21557$ is irreducible over R. In J_2 this polynomial is equal to $x^2 + x + 1$ since $27 \equiv 1$, $159 \equiv 1$, $-21557 \equiv 1 \pmod{2}$. But we have already observed that $x^2 + x + 1$ has no zeros in J_2. Hence $x^2 + x + 1$ is irreducible over J_2 and thus also irreducible over R.

As a more complicated example let us ask whether or not $27x^5 + 40x^3 - 16x^3 - 19x^2 + 100x + 45$ can be factored over R. In J_2 this polynomial is equal to $x^5 + x^2 + 1$. Now obviously, if $x^5 + x^2 + 1$ factors, one of the factors must be linear or quadratic since a cubic factor will produce another factor that is quadratic, and a quartic factor will produce another factor that is linear. As before we observe that the only possible linear factors in J_2 are x and $x + 1$. But $x^5 + x^2 + 1 = x(x^4 + x) + 1 = (x + 1)(x^4 + x^3 + x^2) + 1$ in J_2 so that $x^5 + x^2 + 1$ is not divisible by x or $x + 1$ in J_2. Furthermore, the only possible quadratic factors in J_2 are x^2, $x^2 + x = x(x + 1)$, $x^2 + 1 \equiv (x + 1)(x + 1)$, and $x^2 + x + 1$. We do not need to consider the first three possibilities, however, since, for example, if $x^5 + x^2 + 1$ is divisible by x^2 in J_2 it is certainly divisible by x in J_2—contrary to what we have just shown. Finally, since $x^5 + x^2 + 1 = (x^2 + x + 1)(x^3 + x^2) + 1$ in J_2, it follows that $x^5 + x^2 + 1$ is not divisible by $x^2 + x + 1$ in J_2. (In fact, if we observe that $x^5 + x^2 + 1 = x^2(x + 1)(x^2 + x + 1) + 1$ in J_2 we may eliminate all possible factors immediately.) Thus $x^5 + x^2 + 1$ cannot be factored over J_2 and hence $27x^5 + 40x^4 - 16x^3 - 19x^2 + 100x + 45$ cannot be factored over R.

While irreducibility over J_p implies irreducibility over the rational numbers the converse is not true. Thus $x^2 + 1$ is irreducible over the rational numbers but reducible over J_2 since $x^2 + 1 = (x + 1)(x + 1)$ in J_2. Similarly, $f(x) = 25x^4 + 15x^3 + 21x^2 - 65$ is equal to $x^4 + x^3 + x^2 + 1 = (x + 1)(x^3 + x + 1)$ in J_2. Hence $f(x)$ is reducible over J_2 but we still do not know if it is reducible or irreducible over R. However, we may conclude that if $f(x)$ is reducible over R, $f(x)$ has a *linear* factor. For if $f(x) = g(x)h(x)$ over R where $f(x)$ and $g(x)$ are both quadratic, $f(x)$ will also factor into two quadratic factors over J_2. But since $f(x) = (x + 1)(x^3 + x + 1)$ in J_2, $f(x)$ can factor into quadratic factors in J_2 if and only if $x^3 + x + 1$ is reducible in J_2. But $x^3 + x + 1 = x(x^2 + 1) + 1 = x(x + 1)(x + 1) + 1$ in J_2 so that $x^3 + x + 1$ is irreducible in J_2 since it has a remainder of 1 when divided by x or $x + 1$ (the only two linear polynomials in J_2). Having established that, if $f(x)$ has a factor, it must have a linear factor, we go to J_3 and observe that, in J_3, the only possible linear factors are x, $x + 1$, and $x + 2$ $(= x - 1)$. But $f(x) = x^4 - 2 = x^4 + 1$ in J_3 and hence $f(x) = x \cdot x^3 + 1 = (x + 1)(x^3 - x^2 + x - 1) + 2 = (x - 1)(x^3 + x^2 + x + 1) + 2$ in J_3. Thus $f(x)$ is not divisible by x, $x + 1$, or $x - 1$ and is irreducible over J_3. Hence $f(x)$ is irreducible over R.

The last example serves to bring out another point. Our analysis in J_2 showed that if $f(x)$ had any factors at all it had to have a linear factor. Hence even if our results in J_3 had not been conclusive we would still have known that if we tried all possible linear factors of $f(x)$ over R, (namely, $25x \pm 1$, $25x \pm 13$, $5x \pm 1$, $5x \pm 13$, $x \pm 65$, $x \pm 5$, $x \pm 13$) and none of them were actually factors, there would be no point in looking for a factorization of $f(x)$ into irreducible quadratic factors over R. On the other hand, a test over J_2 easily shows that $g(x) = x^4 + 2x^3 + 3x^2 + 2x + 1$ has no linear factors over J_2 and hence none over R. Thus our search for factors of $g(x)$ over the rational numbers may be confined to a search for quadratic factors and we note that, actually, $g(x) = (x^2 + x + 1)^2$.

Exercises 12.4

1. Find all of the rational zeros of the following polynomials:
(a) $x^4 - x^3 - 7x^2 + x + 6$ (b) $x^3 - 2x^2 + x - 2$
(c) $2x^3 + x^2 - 4x + 6$ (d) $2x^3 - 5x^2 - 10x + 25$
(e) $2x^3 - x^2 - 6x - 10$ (f) $6x^3 - 5x^2 - 4x - 3$

2. Express the polynomials in problems 1(a), (b), (d), and (f) as the product of irreducible factors in $R[x]$; in $R^*[x]$; and in $C[x]$.

3. Express $x^6 - 1$ as the product of irreducible factors in $R^*[x]$ and in $C[x]$.

4. In $J_3[x]$, express the following polynomials as the product of irreducible factors.
(a) $x^2 - x + 1$ (b) $2x^3 - 2x^2 + 1$
(c) $x^4 + x^2 + x + 1$ (d) $x^5 + 1$

5. In $J_5[x]$, express the following polynomials as the product of irreducible factors.
(a) $x^2 - x + 3$ (b) $x^3 + 3x^2 + x - 4$
(c) $2x^3 + 3x^2 + 3x + 1$ (d) $x^4 + 1$

6. Prove that if a, b, and c are all odd integers, then $ax^2 + bx + c = 0$ has no rational roots.

7. Find the g.c.d. of $2x^4 + 9x^2 + 17x - 21$ and $x^3 + 2x^2 + 4x + 21$. Hence find the irreducible factors of the first polynomial in $R^*[x]$ and in $C[x]$.

8. In $C[x]$ find the zeros of the following polynomials and express them in the form $a + bi$ where a and b are real numbers.
(a) $x^2 + ix + 1$ (b) $x^2 + \sqrt{2}x + 2i + 2$

9. Prove Theorem 10.

10. Complete the proof of Theorem 11.

11. Prove the corollary to Theorem 14.

***12.** Prove that in any ring R the following properties are equivalent:
(a) R is a commutative ring and contains no proper divisors of zero;
(b) Every polynomial of degree n with coefficients in R has at most n zeros in R;
(c) Every linear polynomial with coefficients in R has at most one zero in R;
(d) If $a, b, c \in R$ and $b \neq 0$, then $ba = cb$ implies $a = c$.

5. The Cubic and Quartic Equations

The solution of the general quadratic equation $ax^2 + bx + c = 0$ $(a, b, c \in C, a \neq 0)$ was known to the Hindus and the Arabs as early as A.D. 600 (except that they did not recognize the existence of complex numbers). The analogous solutions of the general cubic and quartic equations were developed by the Renaissance Italian mathematicians Tartaglia (1535) and Ferrari (1545) respectively.

To solve the general cubic equation

$$ax^3 + bx^2 + cx + d = 0 \; (a, b, c, d \in C, a \neq 0) \tag{1}$$

we begin by dividing through by a and then replacing x by $y - b/3a$. We leave it to the student to show that we then obtain

$$y^3 + py + q = 0 \; (p = c/a - b^2/3a^2, q = d/a - bc/3a^2 + 2b^3/27a^3) \tag{2}$$

Then (2) is an equation whose roots differ from those of (1) by $b/3a$.

Now we make the substitution $y = z - p/3z$ which, as the student should show, gives us

$$z^3 - p^3/27z^3 + q = 0 \tag{3}$$

Multiplying through by z^3 we obtain

$$z^6 + qz^3 - p^3/27 = 0 \tag{4}$$

which is a quadratic in z^3 whose solutions are given by

$$z^3 = -q/2 \pm \sqrt{q^2/4 + p^3/27} \tag{5}$$

Now we know (Theorem 10.5) that every complex number has three cube roots. Hence, taking $-q/2 + \sqrt{q^2/4 + p^3/27}$ and $-q/2 - \sqrt{q^2/4 + p^3/27}$ in turn, we obtain six values of z. We also know, however, that (2) has at most three different solutions (Theorem 9) so that when we obtain y from $z - p/3z$ we get at most three different pairs of solutions for y, paired solutions being equal.

Example 1. Solve the cubic equation $x^3 + 6x + 2 = 0$.

Solution. The equation is already in the reduced form (2) with $p = 6$ and $q = 2$. Hence we obtain for equation (4), $z^6 + 2z^3 - 8 = 0$ which gives us $z^3 = -4$ or $z^3 = 2$. We now apply De Moivre's theorem and observe that if $\omega = -\frac{1}{2} + (\sqrt{3}/2)i$, the roots of $z^6 + 2z^3 - 8 = 0$ are

$z_1 = \sqrt[3]{-4}$, $z_2 = \sqrt[3]{-4},\omega$ $z_3 = \sqrt[3]{-4}\omega^2$, $z_4 = \sqrt[3]{2}$, $z_5 = \sqrt[3]{2}\omega$, and $z_6 = \sqrt[3]{2}\omega^2$. Using $x = z - 2/z$ we obtain

$$x_1 = z_1 - 2/z_1 = \sqrt[3]{-4} - 2/\sqrt[3]{-4} = \sqrt[3]{-4} + \frac{\sqrt[3]{-8}}{\sqrt[3]{-4}} = \sqrt[3]{-4} + \sqrt[3]{2}$$

$$x_2 = z_2 - 2/z_2 = \sqrt[3]{-4}\omega - 2/\sqrt[3]{-4}\omega$$

$$= \sqrt[3]{-4}\omega + \frac{\sqrt[3]{-8\omega^3}}{\sqrt[3]{-4}\omega} = \sqrt[3]{-4}\omega + \sqrt[3]{2}\omega^2$$

(since $\omega^3 = 1$) and

$$x_3 = z_3 - 2/z_3 = \sqrt[3]{-4}\omega^2 - 2/\sqrt[3]{-4}\omega^2$$

$$= \sqrt[3]{-4}\omega^2 + \frac{\sqrt[3]{-8\omega^3}}{\sqrt[3]{-4}\omega^2} = \sqrt[3]{-4}\omega^2 + \sqrt[3]{2}\omega$$

We leave it to the student to show that the other solutions for z yield no new values for x.

For the quartic equation

$$ax^4 + bx^3 + cx^2 + dx + e = 0 \ (a, b, c, d, e \in C, a \neq 0) \tag{6}$$

we again divide through by a and obtain

$$x^4 + px^3 + qx^2 + rx + s = 0. \tag{7}$$

Our procedure now is to seek to determine numbers α, β, and γ such that

$$x^4 + px^3 + qx^2 + rx + s + (\alpha x + \beta)^2 = [x^2 + (p/2)x + \gamma]^2 \tag{8}$$

for then, by (7), we will have

$$(\alpha x + \beta)^2 = [x^2 + (p/2) + \gamma]^2 \tag{9}$$

and can easily solve for x.

We leave it to the student to show that the conditions for (8) to hold are

$$\alpha^2 + q = 2\gamma + p^2/4, \tag{10}$$

$$2\alpha\beta + r = \gamma p, \tag{11}$$

and

$$\beta^2 + s = \gamma^2. \tag{12}$$

Hence $(\gamma p - r)^2 = 4\alpha^2\beta^2 = 4(2\gamma + p^2/4 - q)(\gamma^2 - s)$ or

$$\gamma^3 - (q/2)\gamma^2 + \tfrac{1}{4}(pr - 4s)\gamma + \tfrac{1}{8}(4qs - p^2s - r^2) = 0. \tag{13}$$

Now we find γ by solving (13) (called the *resolvent cubic equation*) and then can find α and β by substituting in (10) and (11). (Any root of the resolvent cubic will suffice.)

Then (9) gives us

$$x^2 + (p/2)x + \gamma = \alpha x + \beta \tag{14}$$

or

$$x^2 + (p/2)x + \gamma = -\alpha x - \beta \tag{15}$$

so that the four roots of (6) can be found by solving the quadratic equations (14) and (15).

Example 2. Solve the equation $x^4 - 2x^3 - 12x^2 + 10x + 3 = 0$.

Solution. The equations (10), (11), (12), and (13) become, respectively,

(a) $\alpha^2 - 12 = 2\gamma + 1$,

(b) $2\alpha\beta + 10 = -2\gamma$,

(c) $\beta^2 + 3 = \gamma^2$,

and

(d) $\gamma^3 + 6\gamma^2 - 8\gamma - 32 = 0$

Our resolvent cubic, (d), has (fortunately!) the rational root -2. This gives us, by (a) and (b), $\alpha = 3$ and $\beta = -1$. Then (9) becomes $(3x - 1)^2 = (x^2 - x - 2)^2$ so that $x^2 - x - 2 = 3x - 1$ or $x^2 - x - 2 = -(3x - 1)$. Solving these two quadratic equations we obtain as our solution set for the given equation, $\{1, -3, 2 + \sqrt{5}, 2 - \sqrt{5}\}$.

It is obvious that, except for very special values of the coefficients, the solution of cubic and quartic equations by the methods just presented is very difficult. In practice, when the roots cannot be found by simple methods (such as rational roots by the use of Theorem 14 or by De Moivre's theorem in the case of equations of the form $x^n - a = 0$), we are interested mainly in finding the roots approximately to a prescribed degree of accuracy. Many such methods of approximate solution for both real and complex, nonreal, roots can be found in texts on the theory of equations and numerical methods of mathematics.

From both the theoretical and historical point of view, however, the existence of a solution "by radicals" for the linear, quadratic, cubic, and quartic equations is of considerable interest. After the solution of the quartic was obtained it was naturally conjectured that a similar solution existed for the quintic and, even more generally, for polynomial equations of any degree. It was one of the great achievements of nineteenth-century mathematics to discover that no such solution exists for polynomial equations with coefficients complex numbers and of degree ≥ 5. Many of the concepts concerning groups and fields that we have considered here were developed in connection with work on this problem.

It is beyond the scope of this book to give the details concerning this problem here but we would like at least to state the problem and the conclusions precisely. First let us define what we mean by "a solution by radicals."

Definition 9. *The polynomial equation* $f(x) = a_n x^n + a_{n-1}x^{n-1} + \cdots + a_0 = 0$ $(a_i \in C$ *for* $i = 0, 1, \ldots, n;\ a_n \neq 0)$ *is said to be* ***solvable by radicals*** *in C if the roots of* $f(x) = 0$ *can be expressed in terms of the coefficients of* $a_0, a_1, \ldots, a_n$ *using a finite number of additions, subtractions, multiplications, divisions, and extraction of roots.*

We have seen that polynomial equations of degree ≤ 4 are solvable by radicals; the fundamental result of Galois (1811–1832) was that equations of degree ≥ 5 are not solvable by radicals. It is important to note the following points:

(1) We know by Theorem 10.3 that a polynomial equation of any degree with coefficients in C has a solution in C; the result of Galois simply says that the solution cannot, in general, be obtained in a certain form if the degree of the equation is ≥ 5.

(2) Many special equations of higher degree are certainly solvable by radicals (e.g. $x^5 - 1 = 0$); what the result of Galois asserts is that there exists no solution in terms of radicals for the general polynomial equation of degree ≥ 5.

(3) There are standard methods for finding approximate solutions to any polynomial equation; the result of Galois has reference to exact solutions of a certain form.

(4) We are restricting ourselves to a finite number of arithmetic operations. Thus, for example, the use of infinite series is not allowed.

Exercises 12.5

1. Solve the following cubic equations by the method outlined in this section.
(a) $x^3 - 18x + 35 = 0$ (b) $x^3 + 6x + 2 = 0$
(c) $x^3 + 6x^2 + 3x + 18 = 0$ (d) $28x^3 + 9x^2 - 1 = 0$. (*Hint:* First let $x = 1/y$.)

2. Solve the following quartic equations by the method outlined in this section.
(a) $x^4 + 2x^3 - 12x^2 - 10x + 3 = 0$ (b) $x^4 - 8x^3 + 9x^2 + 8x - 10 = 0$
(c) $x^4 - 3x^2 + 6x - 2 = 0$ (d) $x^4 - 10x^2 - 20x - 16 = 0$

3. Verify the results given in equations (2) and (3).

4. Verify that (10), (11), and (12) are the conditions for (8) to hold and that (13) follows from (10), (11), and (12).

PROBLEMS FOR FURTHER INVESTIGATION

1. Investigate the theory of polynomials over an integral domain; over a commutative ring; over a ring. [M2], [V1]
2. Define the derivative of a polynomial in a purely algebraic fashion and derive from this definition the rules for differentiating $f(x) + g(x)$, $f(x)g(x)$, and $[f(x)]^n$ ($f(x)$ and $g(x)$ polynomials, $n \in N$). Investigate the relation of $f'(x)$ to $f(x)$ when multiple zeros of $f(x)$ exist. [W1], [V1]
3. It is well known that if the roots of the quadratic equation $ax^2 + bx + c = 0$ are r_1 and r_2, then $r_1 + r_2 = -b/a$ and $r_1r_2 = c/a$. Generalize these results to polynomial equations of arbitrary degree. [W1]
4. Generalize the notion of the discriminant of a quadratic equation to cubic and quartic equations. [D2]

REFERENCES FOR FURTHER READING

[B4], [C1], [D2], [H4], [M2], [N1], [V1], [W2].

Bibliography

A1 Albert, A. A. *Structure of Algebras.* Providence, R. I. American Mathematical Society, 1939.

A2 Albert, A. A. (editor). *Studies in Modern Algebra* (Vol. 2 of *Studies in Mathematics*). Buffalo, N.Y.: The Mathematical Association of America, 1963.

A3 Albert, R. G. "A Paradox Relating to Mathematical Induction." *American Mathematical Monthly,* **57** (1950): 31–32.

B1 Barnett, R. A. and Fujii, J. N. *Vectors.* New York: John Wiley and Sons, 1963.

B2 Beaumont, R. A. "Equivalent Properties of a Ring." *American Mathematical Monthly,* **57** (1950): 183.

B3 Beaumont, R. A. and Pierce, R. S. *The Algebraic Foundations of Mathematics.* Reading, Mass.: Addison–Wesley Publishing Co., 1963.

B4 Birkhoff, G. and MacLane, S. *A Survey of Modern Algebra* (revised edition). New York: The Macmillan Co., 1953.

B5 Brand, L. "The Roots of a Quaternion." *American Mathematical Monthly.* **49** (1942): 519–520.

C1 Campbell, J. G. "Another Solution of the Cubic Equation." *Mathematics Magazine,* **35** (1962): 43.

C2 Carmichael, R. D. *Introduction to the Theory of Groups of Finite Order.* New York: Dover Publishing Co., 1956.

C3 Curtis, C. W. and Reiner, I. *Representation Theory of Finite Groups and Associative Algebras.* New York: John Wiley and Sons, 1962.

D1 Dean, R. A. "Group Theory for School Mathematics." *The Mathematics Teacher.* **LV** (1962): 98–105.

D2 Dickson, L. E. *First Course in the Theory of Equations.* New York: John Wiley and Sons, 1922.

D3 Dubisch, R. "The Wedderburn Structure Theorems." *American Mathematical Monthly,* **54** (1947): 253–259.

D4 Dubisch, R. "Representation of the Integers by Positive Integers." *American Mathematical Monthly,* **58** (1951): 615–616.

D5 Dubisch, R. *The Nature of Number.* New York: The Ronald Press Co., 1952.

D6 Dubisch, R. *Trigonometry.* New York: The Ronald Press Co., 1955.

D7 Dubisch, R. "A Chain of Cyclic Groups." *American Mathematical Monthly,* **66** (1959): 384–386.

E1 Ellis, W. "A Nonmodular Field: Constructive Approach." *The Mathematics Teacher.* **LV** (1962): 544–548.

E2 Eves, H. and Newsom, C. V. *An Introduction to the Foundations and Fundamental Concepts of Mathematics.* New York: Holt, Rinehart, and Winston Co., 1958.

F1 Fletcher, T. J. "Campanological Groups." *American Mathematical Monthly.* **63** (1956): 619–626.

G1 Golomb, S. W. "Distinct Elements in Non-Commutative Groups and Loops." *American Mathematical Monthly*, **70** (1963): 541–544.

G2 Guy, W. T., Jr. "On Equivalence Relations." *American Mathematical Monthly*, **62** (1955): 179–180.

H1 Halmos, P. R. *Finite Dimensional Vector Spaces* (2nd edition). Princeton, N.J.: D. Van Nostrand Co., Inc., 1958.

H2 Halmos, P. R. *Naive Set Theory*. Princeton, N.J.: D. Van Nostrand Co., Inc., 1960.

H3 Hamilton, N. and Landin, J. *Set Theory, The Structure of Arithmetic*. Boston: Allyn and Bacon, Inc., 1961.

H4 Hellman, M. J. "A Unifying Technique for the Solution of the Quadratic, Cubic, and Quartic." *American Mathematical Monthly*, **65** (1958): 274–276.

H5 Hensel, K. *Zahlentheorie*. Berlin: Goschen, 1913.

H6 Herstein, J. N. *Topics in Algebra*. New York: Blaisdell Publishing Co., 1964.

J1 Jacobson, N. *The Theory of Rings*. Providence, R. I.: American Mathematical Society, 1943.

J2 Johnson, H. H. "On the Anticenter of a Group." *American Mathematical Monthly*, **68** (1961): 469–472.

J3 Jones, B. W. *The Arithmetic Theory of Quadratic Forms* (Carus Monograph No. 10). Buffalo, N.Y.: The Mathematical Association of America, 1950.

K1 Kamke, E. *Theory of Sets*. New York: Dover Publishing Co., 1950.

K2 Kelley, J. F. *Introduction to Modern Algebra*. Princeton, N.J.: D. Van Nostrand Co., Inc., 1960.

K3 Kershner, R. B. and Wilcox, L. R. *The Anatomy of Mathematics*. The Ronald Press Co., 1950.

K4 Kurosh, A. G. *The Theory of Groups*, Vol. I. New York: Chelsea Publishing Co., 1956.

L1 Landau, E. *Foundations of Analysis*. New York: Chelsea Publishing Co., 1951.

L2 Levine, N. "The Anticenter of a Group." *American Mathematical Monthly*, **67** (1960): 61–63.

L3 Luh, J. "Ideals and Adeals of a Ring." *American Mathematical Monthly*, **70** (1963): 548–550.

M1 Maria, M. H. *The Structure of Algebra and Arithmetic*. New York: John Wiley and Sons, 1958.

M2 McCoy, N. H. *Rings and Ideals* (Carus Mathematical Monograph No. 8). Buffalo, N.Y.: The Mathematical Association of America, 1948.

M3 McCoy, N. H. *Introduction to Modern Algebra*. Boston: Allyn and Bacon, Inc., 1960.

M4 Miser, W. L. "Summing Series Whose General Terms are Polynomials." *Mathematics Magazine*, **7** (1931–32): 17–19.

M5 Mulcrone, T. F. "Semigroup Examples in Introductory Modern Algebra." *American Mathematical Monthly*, **69** (1962): 296–301.

M6 Murdoch, D. C. *Linear Algebra for the Undergraduate*. New York: John Wiley and Sons, 1957.

N1 Nagle, T. D. "A Method for the Solution of the General Cubic." *American Mathematical Monthly*, **59** (1952): 326–327.

N2 Niven, I. "Equations in Quaternions." *American Mathematical Monthly*, **48** (1941): 654–661.

N3 Niven, I. and Zuckerman, H. S. *An Introduction to the Theory of Numbers.* New York: John Wiley and Sons, 1960.

O1 Ore, O. *Number Theory and Its History.* New York: McGraw–Hill Book Co. 1948.

P1 Pollard, H. *The Theory of Algebraic Numbers* (Carus Monograph No. 9). New York: John Wiley and Sons, 1950.

R1 Roberts, J. B. *The Real Number System in an Algebraic Setting.* San Francisco: W. H. Freeman and Co., 1962.

R2 Rosenbaum, R. A. "Remark on Equivalence Relations." *American Mathematical Monthly*, **62** (1955): 650.

S1 Sholander, M. "Postulates for Commutative Groups." *American Mathematical Monthly*, **66** (1959): 93–95.

S2 Slater, M. "A Single Postulate for a Group." *American Mathematical Monthly*, **68** (1961): 346–347.

S3 Stoll, R. L. *Sets, Logic, and Axiomatic Theories.* San Francisco: W. H. Freeman and Co., 1961.

T1 Tarski, A. *Introduction to Logic and to the Methodology of Deductive Science.* New York: Oxford University Press, 1941.

T2 Thurston, H. S. *The Number System.* New York: Interscience Publishing Co., 1956.

U1 Utz, W. R. "Square Roots in Groups." *American Mathematical Monthly*, **60** (1953): 185–186.

V1 Van der Waerden, B. L. *Modern Algebra*, Vol. 1 (second edition). New York: F. Ungar Publishing Co., 1953.

W1 Weiss, M. J. *Higher Algebra for the Undergraduate* (second edition). New York: John Wiley and Sons, 1962.

W2 White, C. R. "Definitive Solutions of the General Quartic and Cubic Equations." *American Mathematical Monthly*, **69** (1962): 285–287.

W3 Whittaker, J. V. "On the Postulates Defining a Group." *American Mathematical Monthly*, **62** (1955): 636–640.

Z1 Zassenhaus, H. J. *The Theory of Groups* (second edition). New York: Chelsea Publishing Co., 1958.

Index